AF419910

Natural Products:
A Practical Manual

Natural Products:
A Practical Manual

Dr P. Praveen Kumar

M.Pharm Ph.D

Department of Pharmaceutical Chemistry

Hindu College of Pharmacy, Guntur, A.P

PharmaMed Press

An imprint of Pharma Book Syndicate

An unit of **BSP Books Pvt., Ltd.**

4-4-309/316, Giriraj Lane,

Sultan Bazar, Hyderabad - 500 095.

Fourth Reprint 2019

Published by :

PharmaMed Press
An imprint of Pharma Book Syndicate
An unit of BSP Books Pvt., Ltd.
4-4-309/316, Giriraj Lane, Sultan Bazar, Hyderabad - 500 095.
Phone: 040-23445600, 23445688; Fax: 91+40-23445611
E-mail: info@pharmamedpress.com
www.pharmamedpress.com/pharmamedpress.net

ISBN : 978-93-89974-86-7

Message

Dr. Mannava Radha Krishna Murthy
MD
Chairman, Hindu College of Pharmacy,
EC member, Acharya Nagarjuna University,
Guntur.

I am extremely glad and proud to know that Mr. Praveen Kumar is bringing out the book Natural Products: A Practical manual.

I congratulate Mr. P. Praveen Kumar, who is working as an Assistant Professor in our Hindu College of Pharmacy, Guntur for his dedication, commitment and taking pain in the completion of this manual inspite of his busy schedule in the college.

I hope this book will fulfil all the expectations and provide the necessary requirements of undergraduate and postgraduate pharmacy students.

Dr. Mannava Radha Krishna Murthy

Foreword

Dr.Y.Rajendra Prasad,
M.Pharm, Ph.D.,
University College of Pharmaceutical sciences,
Department of pharmaceutical chemistry,
Andhra University, Visakhapatnam.

Mr. P. Praveen Kumar has painstakingly compiled the Natural Products: A practical manual a very useful book for undergraduate and postgraduate pharmacy students. This book covers various dimensions of the entire curriculum on natural products of medicinal interest. I congratulate the author for bringing out this much wanted book and I strongly recommend this book to pharmacy students.

Dr.Y.Rajendra Prasad

Preface

The study of natural products has always been the starting point of the discipline of Chemistry in every country of the globe, in view of the importance of the organic compounds in agriculture, medicine and industry. Every student of chemistry today feels the need to acquire further knowledge in this field. This laboratory manual is designed to meet the needs in laboratory experiments for natural products, and phytochemistry, as such, this book will be considered by faculty, undergraduates and postgraduates of pharmacy courses of various universities.

Each chapter includes a general introduction and method of isolation, degradation and applications of chromatographic procedures. The introduction in each chapter is brief and attempts only to supply or recall knowledge in the particular field. The student, who does not always find the time to read the relevant books or reviews, will find, in this introduction, the required material in a concentrated form.

Each experiment is described under the following headings :

Introduction, principle, materials, procedures and chemical tests (often including spectral data)

I express my profound sense of gratitude and indebted to Dr. Y. Rajendra Prasad M.Pharm, Ph.D, Department of Pharmaceutical Chemistry, University College of Pharmaceutical Sciences, Andhra University, Visakhapatnam for his encouragement, constructive comments and criticism.

I am deeply indebted to our Chairman, Dr. Mannava Radha Krishna Murthy and Secretary, G. Godavarthy Sathyanarayana, and Committee members of Hindu College of Pharmacy, Guntur for their continuous co-operation and encouragement.

I acknowledge with thanks to all my friends and well wishers for their encouragement rendered directly or indirectly at all times in the completion of this practical manual.

Acknowledgements are also due to Mr. S. Sridhar Raju, Lecturer, Chips, Guntur for editing the manuscript and drawing structures.

Thanks are due to our publisher, PharmaMed Press, Hyderabad, for publishing this book in a satisfying manner.

Last, but certainly not least, I wish to thank my wife P. Anupama for her interest and encouragement.

The author is hopeful that the book will fulfil the expectations of degree and postgraduate students of Pharmacy. I have made all efforts to make it an ideal book; However I shall look forward to valuable suggestions for the improvement of the book.

-Author

Contents

Chapter - 7

Vitamins

Chapter - 8

Analysis of Fixed Oils or Fats

Chapter - 9

Estimaton of Nitrogen (Kjeldahl's Method)

Chapter - 10

Estimation of Carbonyl Group

Chapter - 11

Estimation of Hydroxyl Group

Chapter - 12

Interpretation of IR Spectroscopy

Chapter - 13

1

Carbohydrates

1.1 Introduction

Carbohydrates are widely distributed in plants and animals. They are source of energy and also as store of energy. Certain carbohydrates (cellulose) support the plant tissue, while some other form the major constituent of the (chitin) shells of crabs and lobsten. There are various other important contributions, sugar makes fruit sweet and yield alcohol on fermentation, cellulosic materials such as cotton, linen, jute, straw, grass, wood, supply cotton, paper, fuel, plastics, paints and explosives. Ribose and deoxy ribose are compounds of nucleic acids RNA and DNA,which determine human heredity.

"Carbohydrates are defined as the optically active polyhydroxy aldehydes or ketones or substances, this can be hydrolyzed to either of them"

Carbohydrates are divided into two main groups:

- Sugars
- Non-sugars (Polysaccharides)

Sugars are crystalline substance with a sweet taste and soluble in water.

Sugars are subdivided into two main groups:

- Monosaccharide
- Oligosaccharides

The simplest monosaccharide is glycolaldehyde. This does not contain an asymmetric carbon atom and is therefore not optically active. Since all naturally occurring sugars are optically active.

Hexoses ($C_6H_{12}O_6$)

Aldohexose and ketohexose are important groups of monosaccharide. The Aldohexose (Glucose) contains four structurally different asymmetric carbon atoms, and can therefore exist in sixteen optically active forms. All are known, and correspond to the D and L forms of glucose, mannose, galactose, fructose etc. The mirror images of a sugar (D and L) are called optical antipodes or enantiomers. They have identical chemical and physical properties and differ only in the sign of these optical rotations.

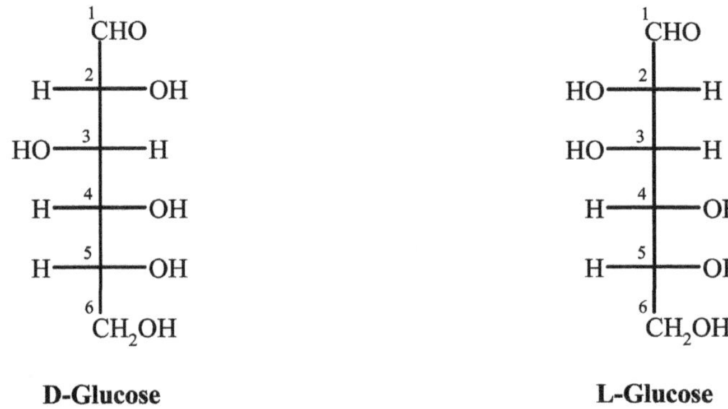

D-Glucose   **L-Glucose**

The structure of glucose and mannose contain four asymmetric identical groups, except for these at C_2, which are opposite in space. Such isomers are called epimers.

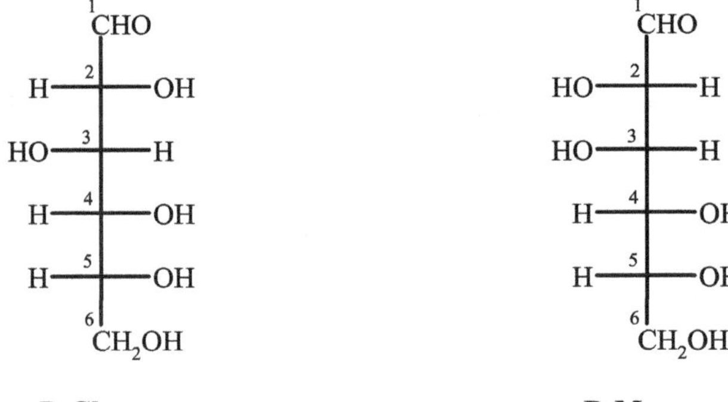

D-Glucose   **D-Mannose**

It has been demonstrated that the free monosaccharide also exist in two forms are structurally related to methyl glucoside.The ring may be five membered – a derivative of tetrahydro furan (furanose) or six membered – a derivatives of pyran (pyranose)

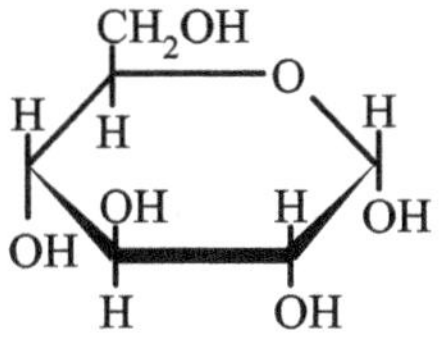

α-D-Glucopyranose

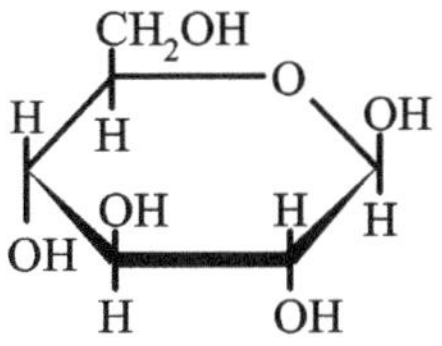

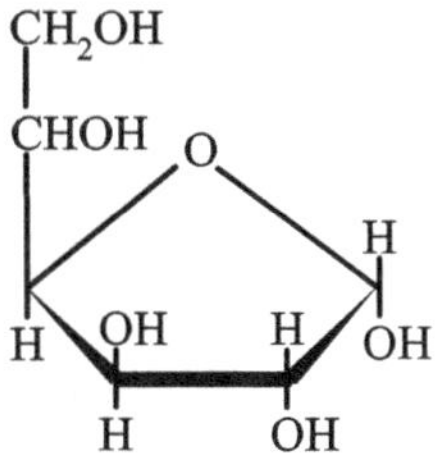

β-D-Glucopyranose

α-D-Glucofuranose

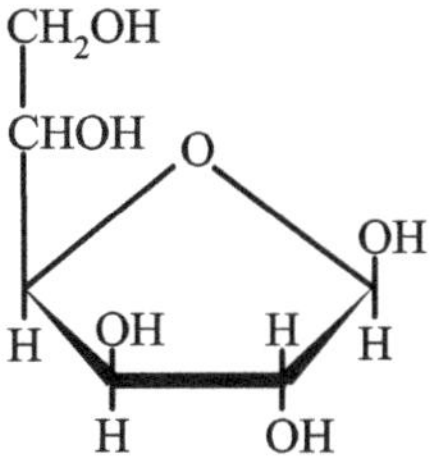

β-D-Glucofuranose

The important ketohexose is Fructose. This contains three structurally different asymmetric carbon atoms and can therefore exist in eight optically active forms. Fructose is found in fruits, honey and combined with glucose in the disaccharides cane sugar from which, it may be obtained by hydrolysis and fractional crystallization. D-Fructose in similar manner, Howarth proposed a five membered ring for furanose and six membered rings for pyranose.

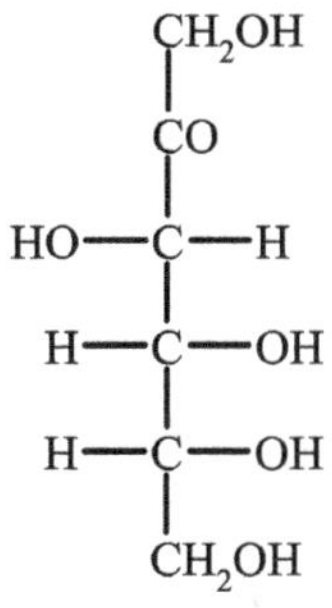

D-Fructose

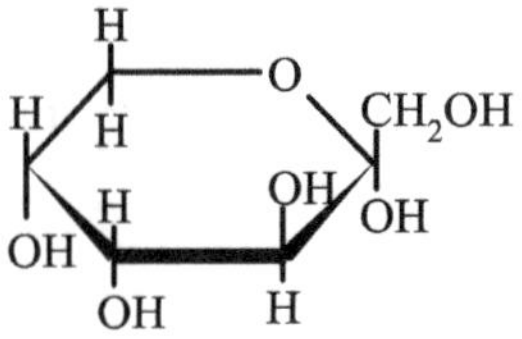

α-D-Fructopyranose

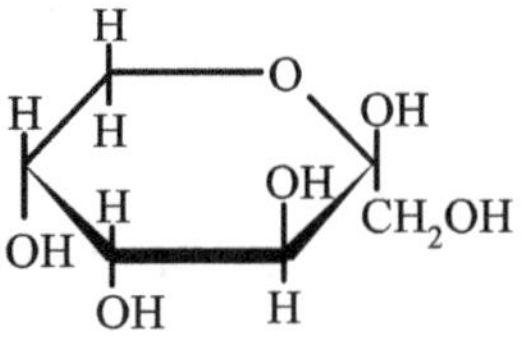

β-D-Fructopyranose

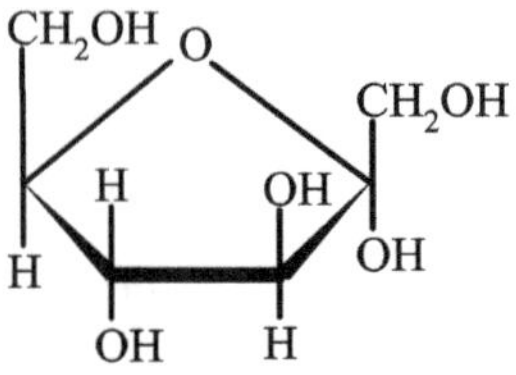

α-D-Fructofuranose

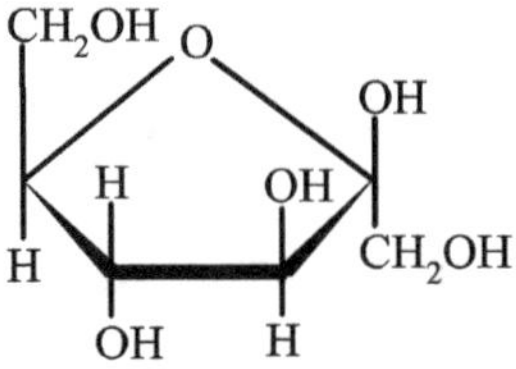

β-D-Fructofuranose

Phenyl hydrazine and other substituted hydrazines are important in the identification of the various monosaccharides. When one mole of phenyl hydrazine reacts with one mole of an aldohexose or ketohexose, the product is a normal hydrazone. On treating with an excess of phenyl hydrazine, an osazone formed. Hydrolysis of the osazone with HCl yields the corresponding Osones, which are polyhydroxy-2-keto-1-aldehyde.

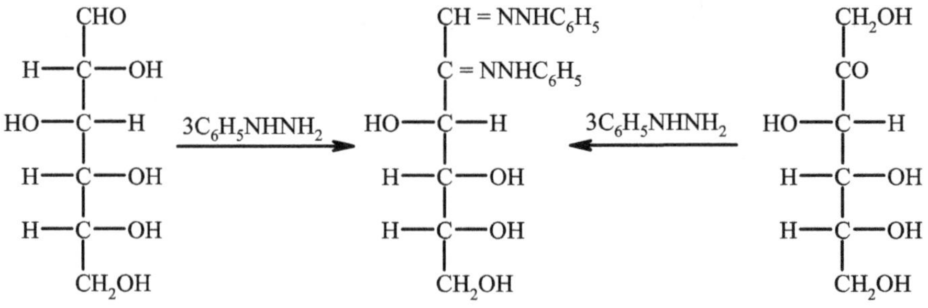

D-Glucose $\xrightarrow{3C_6H_5NHNH_2}$ Osazone $\xleftarrow{3C_6H_5NHNH_2}$ D-Fructose

It should be noted that the osazone formed from glucose, mannose and fructose are identical and hence the lower carbon atoms C_3-C_6 of these sugars have identical configurations.

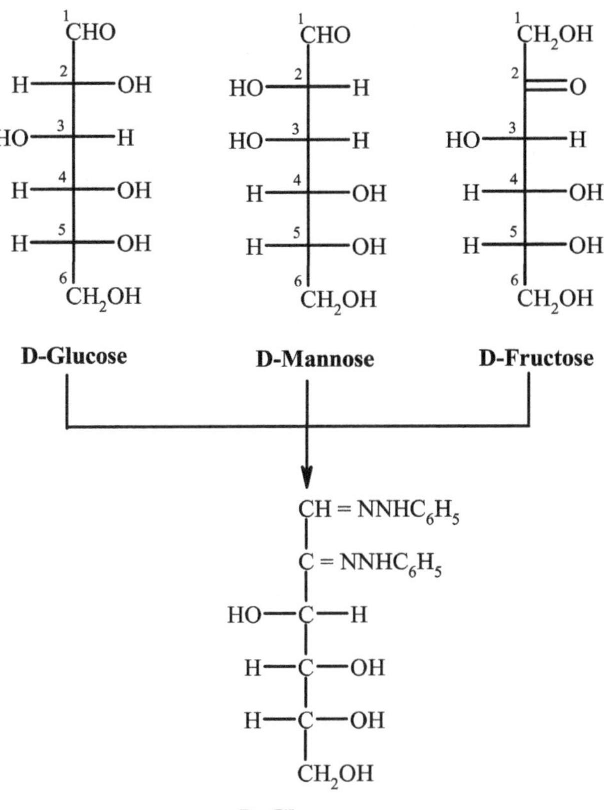

In presence of dilute alkali or amines, sugers undergo rearrangement to form a mixture of more than one sugar. In presence of very dilute alkali (0.02N NaOH) glucose is converted into mixture of glucose, mannose and fructose. Mannose and fructose also undergo the same changes to form a similar mixture. This reaction was called the Lobryde Bruyn-Van Ekenstein conversion. Wohl in 1900 suggested that a common enediol intermediate is formed during the transformation through 1, 2 enolism.

The 1, 2 enolisation steps during the transformation were proved by Topper et. al., (1951). But side by side, he concluded that mannose and fructose are not formed from a single enediol, but two geometrical isomeric enediol intermediates are formed.

CHO
H——OH
HO——H
H——OH
H——OH
CH₂OH
D-Glucose
H—C—OH
——OH
HO——H
H——OH
H——OH
CH₂OH
Enedoil
CHO
HO——H
HO——H
H——OH
H——OH
CH₂OH
D-Mannose
CH₂OH
C=O
HO——H
H——OH
H——OH
CH₂OH
D-Fructose
CHO
H—C—OH
R
D-Glucose
HO—C—H
C—OH
R
Trans-diol
CH₂OH
C=O
R
D-Fructose
HO—C—H
HO—C
R
Cis-diol
CHO
HO—C—H
R
D-Mannose

Mutarotation

When glucose is dissolved in water its specific rotation gradually changes until it reaches a constant value. This phenomenon, known as mutarotation, is shown by almost all the reducing sugars (Except a few ketones).

For e.g., when glucose prepared by crystallization from water below 50 °C is dissolved in water its initial specific rotation of + 111° falls gradually to a constant value of + 52.5°. Similarly when the glucose crystallized from water above 95 °C is dissolved in water its initial specific rotation of + 19.5° gradually rises to + 52.5°.

The ring structure for glucose also explains the mutarotation by the fact that the change in specific rotation is due to the interconversion of the alpha form [$(\alpha)_D = +113°$] of glucose to β form [$(\alpha)_D = + 19°$] and vice versa through the aldehydic structure till an equilibrium is reached between the two structures. The specific rotation value of this equilibrium mixture corresponds to + 52.5°.

α-D-Glucose	Aldehydic form	β-D-Glucose
$(\alpha)_D = + 111°$	$(\alpha)_D = + 52.5°$	$(\alpha)_D = + 19.5°$

Oligosaccharides

These are the sugars composed of two to ten monosaccharide units linked together in a branched or unbranched chain. The oligosaccharides are classified as di, tri, tetra – etc., depending upon the number of monosaccharide produced upon acid hydrolysis.

Disaccharides: In disaccharides, if the linkage through the glycosidic carbon atoms of each component C1-C1, the disaccharide is non-reducing sugar (sucrose and trehalose). If the linkage through C_1 of the first and C_4 or C_6 of the second component, the disaccharide is reducing sugar (cellobiose and gentiobiose).

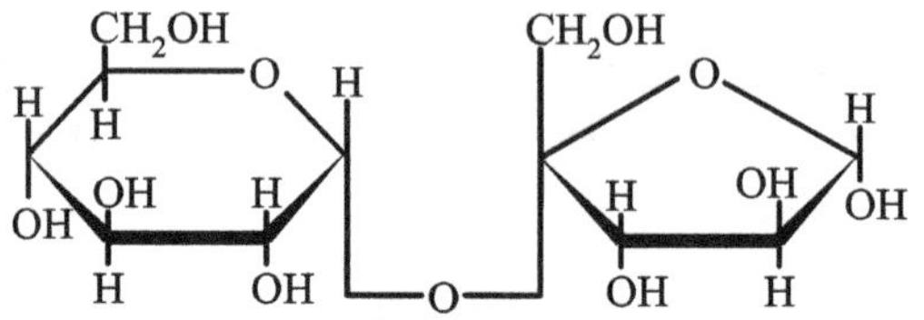

Sucrose

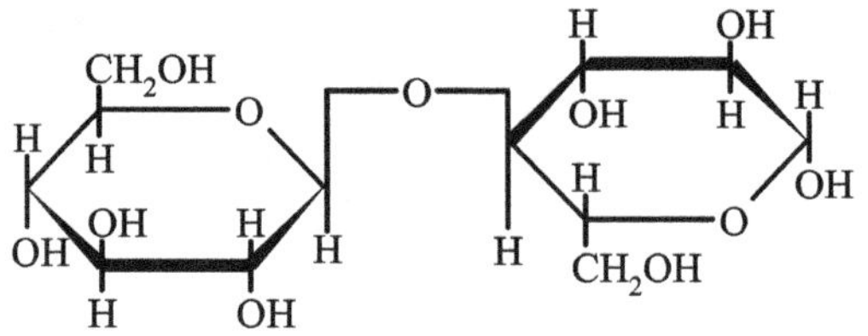

Cellobiose

Raffinose is an example of a naturally occurring trisaccharides

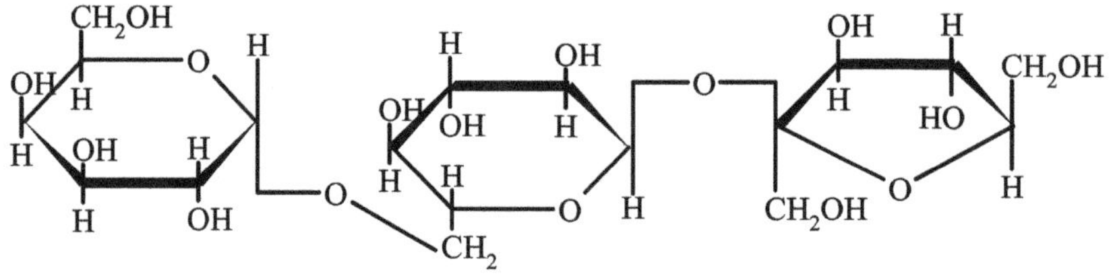

Polysaccharides

Polysaccharides are high molecular weight carbohydrates which yield ultimately monosaccharides on complete hydrolysis and hence chemically, they are regarded as the condensation products of monosaccharides. They are found in the higher plants, in seaweed, in fungi, in bacteria, and in animals, where they serve as structural support and as food reserves.

Polysaccharides classified mainly in two groups:

(i)   *Homopolysaccharides* contains only one kind of monosaccharides units (Cellulose, Inulin and Starch).

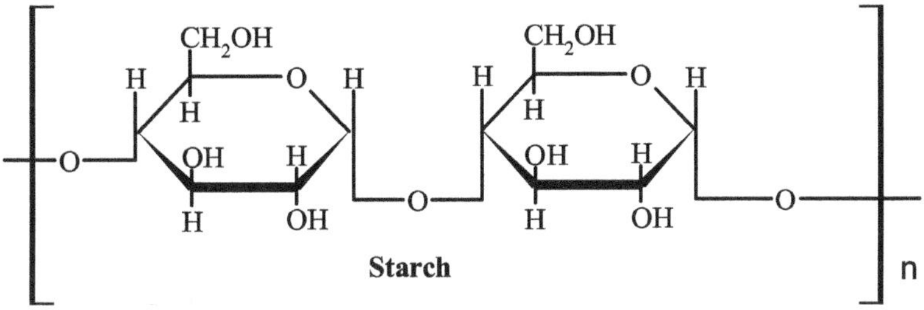

$$CH_2OH \qquad CH_2OH$$

Starch

(ii)   *Heteropolysaccharides* contain more than one kind of monosaccharide units (Heparin).

Reducing and Non Reducing Sugars

Alkaline copper solution readily oxidizes many mono and disaccharides with the oxidation of the sugar. There is a simultaneous reduction of the cupric ion (cu^{+2}) to a cuprous ion (cu^{+1}) compound, which can be identified easily by the color changes from a deep blue to a red or reddish brown.

The best known solutions for these tests are:

(a) Fehling's reagent

(b) Benedict's reagent

(c) Barford's reagent

The following experiments will differentiate between reducing and non reducing sugars. Place 10 ml of benedict's solution in each of six test tubes. Add 50 mg of glucose to the first, fructose to second, starch to the third, sucrose to the forth, lactose to the fifth and in the sixth tube starch with 10 ml saliva. Shake well and allow the tubes to stand for 15 min. After this, place the six test tubes in a 600 ml beaker that is a third full of water and heat it gently for 10-15 min. Record the results.

1.2   Isolation of Starch from Potatoes

Introduction

Starch is a polysaccharide made with the basic units of D-glucose. The pharmaceutically used starch is obtained from the grains of maize (Zea maize) or rice (Oryza sativum) or wheat (Triticum aestivum) or potato tubers (Solanum bnotatum). The starch contains chemically two components.

1. α Amylose which is water insoluble and gives blue with iodine.

2. β amylopectin which is water soluble and gives violet colour with iodine.

The glucose units in the starch are linked by α- 1, 4 and α- 1, 6 glycosidic linkage. The amylase is a linear unit made up of α -1, 4 linkage where as amylopectin cross linked made up 1, 4 and 1, 6 glycosidic linkage (branching γ chain).

Many different methods have been used to fractionate starch into its branched and linear components. The method most suitable for one kind of starch is not necessarily the best for another and much depends upon the source and pretreatment of the starch.

In 1941 Schoch demonstrated that slow cooling of a hot aqueous starch solution saturated with butanol gives a microcrystalline precipitate of the linear starch component, amylase. Fractionation of starch by leaching techniques was applied by Baum and Gilbert, who utilized the insolubility in cold, dilute alkali of the amylopectin fraction from undamaged starch granules.

Materials

Butanol,   Sodium chloride,   HCl,

Potatoes,   Sodium hydroxide

Procedure

1 kg undamaged potatoes are peeled, sliced, and then extracted in 250 g lots with 750 ml 1% sodium chloride solution in a mechanical food blender. Each lot is blended for about half a min, and the slurry is then filtered, through fine muslin cloth with gentle agitation. The filter cakes are bulked and reextracted for one min with 150 ml salt solution. The slurry is filtered through muslin as before, and the filtrates are bulked. The starch granules are allowed to settle, and the supernatant, together with fragments of cellulose that have passed through the muslin, is decanted and discarded. The starch is washed three times with 1% sodium chloride and then once with 0.01N sodium hydroxide. The solution is finally washed three times with distilled water and stored underwater at 4 °C. Precautions must be taken against freezing.

Sources and characteristics of various starches:

Starch type	source	granule shape	granule size (μm)
Corn (maize)	seed	round or polygonal	5-25
Potato	root	egg-shaped	15-100
Sweet potato	root	polygonal	10-25
Wheat	seed	round or elliptical	2-10 or 20-35
Sago	stem	oval or egg-shaped	20-60
Rice	seed	polygonal	3-8
Barley	seed	round or elliptical	2-6 or 20-35

Spectral data:

IR Spectroscopy(cm^{-1}):

-OH (str)	3350
-CH (str)	2929, 2898
$-CH_2$ (bend)	1452
glycosidic C-O-C assymmetric (str)	1148
(ring vibration)	930
(C-1 group vibration)	862
(ring breathing vibration)	764
(low frequency ring vibrations)	709, 643, 606, 576, 524

$^{13}CNMR$ (δppm):

101.2	(C_1)
72.2	(C_4)
81.5	(C_2 & C_3 & C_5)
61.9	(C_6)

1.3   Isolation of Amylopectin and Amylase from Potato Starch

Introduction

Starch is separated into its principal components amylase and amylopectin, by addition of dilute alkali followed by neutralization. Amylase remains in solution, while amylopectin forms a gel. The former is precipitated selectively from the saline solution by addition of n-butanol; removal of the butanol from the amylase-butanol complex yields pure amylase. Amylopectin is isolated by centrifugation followed by freez drying.

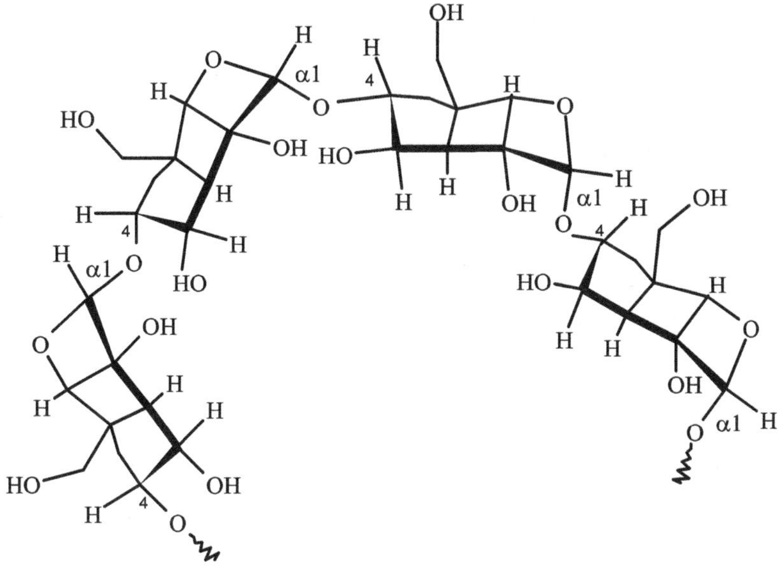

Materials

Freshly prepared starch

Sodium hydroxide (0.157N)

1N HCl

Procedure

To 3.3 liters of 0.157N NaOH in a glass walled tank and added 40 gm of drained, freshly prepared starch and suspended mixture in 260 ml water with gentle shaking. The temperature of the dispersion must be at 22 °C but should preferably be near 25 °C. The mixture should be stirred gently until it clears. The alkaline solution should stand for a total of 5 min. 915 ml 5% NaCl solution is added, and the dispersion is neutralized with 1N HCl to pH 6.5-7.5 with the aid of an external indicator or long lead glass electrode assembly.

After standing over night at room temperature, the gel settles until it occupies about one third of the total volume. There should be a sharp division between the gel and the amylase solution and later is then removed by suction, and should be filtered through a 15 cm fritted glass filter (No. 3).This solution is then filtered through a No.4 filter.

Precipitation of Amylose

The amylose retrogrades, if stirred as a saline solution for more than a few days; it should thus be separated from this solution as soon as possible. Butanol is used to precipitate the amylase selectively from the saline solution according to the following procedure. The filtrate is saturated with redistilled butanol and stirred gently with a motor driver's stirrer for about 1 hr at room temperature. The precipitated amylase butanol complex is allowed to settle for 2-3 hr and the clear supernatant is siphoned off.

The partially sedimented solid is then centrifuged at 3000 rpm for 15 min in polypropylene buckets. The complex is then stirred in butanol saturated water (125 ml) and reconcentrated by centrifuging as before. The complex is stable if stored in a stoppered container. It is readily soluble in water at room temperature. To remove most of the butanol from this solution, oxygen free nitrogen is bubbled through it for 10 min in a vessel heated in a boiling water bath; 40 g wet potato starch yield 1.5-1.85 g amylose.

Purification of Amylopectin

The amylopectin gel is centrifuged at 8000 rpm for 20 min at 20 °C. The supernatant is discarded, and 1% NaOH solution is added to the gel (100 ml per 4 g original wet starch) with stirring, the mixture is allowed to stand for 20 hr at 18 °C, and the gel is then collected by centrifuging at 8000 rpm as before. A second washing with NaCl is necessary. After centrifuging the gel can be either suspended in water for immediate use (insoluble in cold water) or freez-dried and stored; 40 g wet starch yields approximately 6 g amylopectin.

Chemical Test

A few granules of the starch in 1 ml water are treated with several drops of a dilute solution of iodine in KI, it gives intense blue color. Amylose exhibits a blue color, and amylopectin develops reddish color.

Pharmaceutical Use

Starch is used as nutritive, demulcent, protective, disintegrating agent and binding agents (in formulation of tablets). It is used as antidote for iodine poisoning.

1.4   Hydrolysis of Starch and its Qualitative Analysis

Principle and Discussion

Starch is known to be a polysaccharide, made up of D-glucose is basic unit which are linked by glycosidic linkage. The linkages highly sensitive that undergo hydrolytic cleavage in presence of concentrated HCl under endothermic condition or in room temperature:

The starch undergoes hydrolysis in the following pattern

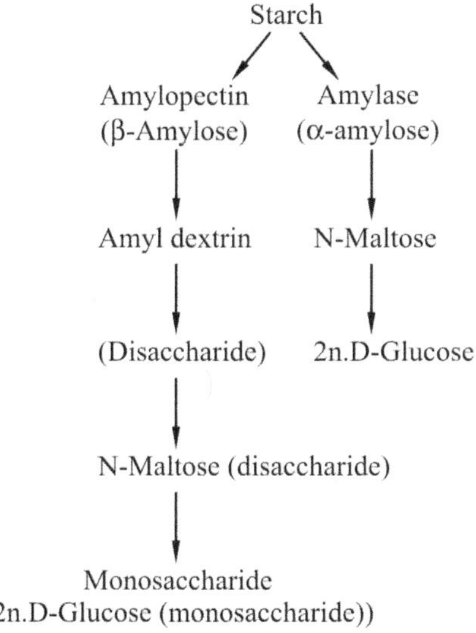

At the end of complete hydrolysis of starch, starch undergoes glycosidic cleavage and a number of maltose followed by production of 2n number of glucose during the hydrolytic pathway of starch. The intermediate is amylodextrin, which is a secondary derivative of amylopectin, which gives red color with iodine, where as amylopectin gives violet color with iodine. The complete hydrolysis of starch yields maltose which can be identified with negative test to iodine. Further, the complete hydrolysis of starch may be supported by carbohydrate analysis for the presence of maltose, glucose, starch and dextrin.

Procedure

To 1 gm of starch add 1 ml of concentrated HCl and 5 ml of water and heat on a water bath at 60-70 °C. An observation was made on solution for presence of starch by iodine test, continued heating until the solution shows negative to iodine test and positive for Benedicts or Fehlings test. Cool the solution to room temperature to carry out systematic carbohydrate analysis for the presence of maltose.

1.5 Isolation of Aloin from Aloe

Introduction

Aloin is a C-glycoside which is having anthraquinone as aglycone and rhamnose (L-configuration) as glycone moiety, this aloin is called as aloinoside which posses two isomers aloniside A (aloin A), aloniside B (aloin B). Aloin is nothing but barbaloin which is a mixture of both diasterioisomers A & B at C10 of aloin amodine anthrone (a glycon). The aloe barbadensis consist of aloinemodine, chrysophenol and aloe resin B in minor quantity.

Indian aloes contain alenoside as major constituent, only with traces of aloin. Barbolin is 10-glucophyranosyl derivative of aloinemoidine which is not hydrolyzed by alkaline or acid hydrolysis but hydrolyzed by oxidative hydrolysis.

The isomers aloin or barbaloin A exhibit S – configuration at C10th carbon, where as B – aloin exhibit R-configuration. Both R & S configurations are inter convertible via anthranol formation.

Procedure

Isolation is done by cutting the leaves near the bases, by which the juice located in parenchymatous cells. After cutting the leaves are placed along the sides of V-shaped wooden troughs. Because of the spines on the leaves they are put into kerosene tins, to drain out of all the juice. The collected juice is boiling in large copper pans. Then the juice is poured into metal containers where it hardens.

Uses

Used as purgative, prolonged use of aloin may effect electrolyte balance (potassium deficiency) and disturb cardiac rhythm. The large doses of aloin stimulate uterine muscle in case of pregnant women.

1.6   Isolation of Pectin from Orange Peel

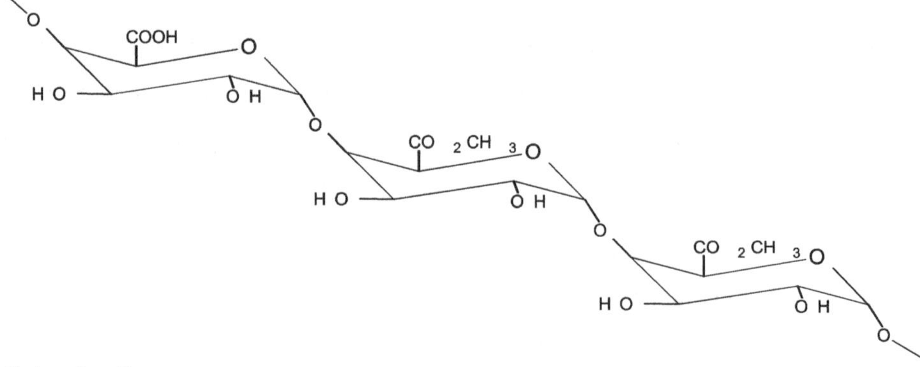

Introduction

Pectins are polygalacturonic acid esters and can have molecular weights from 20,000 to 400,000. These polysaccharides are present in many fruits, particularly in orange and lemon rinds, where they may constitute as much as 30% of the total weight.

The pectin in orange peel is a soluble fiber that may be helpful in lowering blood cholesterol and glucose levels. Soluble fibers are thought to prevent cholesterol absorption and slow the absorption of dietary sugar. The pectin in orange peel appears to stimulate the growth of healthy bacteria that is effective in helping to prevent food-borne pathogens and some folk healers have even claimed that orange rind may also help to clear toxins (including lead poisoning) from the system.

Materials

Fresh orange peel,   Citric acid,   Sulfuric acid,

Ethanol,   Acetone

Procedure

Cut the peels in to small pieces, weigh about 50 g of fresh peel. Wash thoroughly with water and immerse in 200 ml of deminaralised water. Adjust the PH to 4 with the addition of citric acid or tartaric acid or by addition of 3N H_2SO_4. Heat at 85-90°C for about one hour with stirring. Add filter aid and filter immediately, while solution is hot. Cool the filtrate and pour it slowly into 3 volumes of acetic isopropanol or ethanol or acetone, stir the solution thoroughly for precipitation of pectin. Filter through nylon sieve and wash it several times with small volumes of 70 % isopropanol or acetone in order to remove it free from acidic ions. Dry the product in vacuum, weigh and store in well closed container.

1.7   Isolation of Sennosides as Calcium Salts from Senna Leaves

Introduction

Senna was first isolated and characterized by Stoll in 1941. The first two glycosides were identified and attributed to the anthraquinone family. These were found to be dimeric products of aloe emodin and/or rhein which were named sennoside A and sennoside B. They both hydrolyze to give the aglycones sennidin A and B and two molecules of glucose. Senna consists of the dried leaflets or fruits of Cassia senna (C. acutifolia) known in commerce as Alexandrian senna and Cassia angustifolia commonly known as Tinnevelly senna.

The senna plants are small shrubs of Leguminosae cultivated either in Somalia, the Arabian Peninsula or near the Nile River. Tinnevelly senna is obtained from cultivated plants mainly in South India and Pakistan.

Senna contains anthraquinone glycosides known as sennosides. These molecules are converted by the normal bacteria in the colon into rhein-anthrone, which in turn has two effects. It first stimulates colon activity and thus speeds bowel movements. Second, it increases fluid secretion by the colon. Together, these actions work to get a sluggish colon functional again. Senna leaves and pods have been shown to have laxative activity. It is useful in habitual constipation.

Materials

Electric shaker,	Vacuum filter,	Desiccator
Benzene,	Methanol,	HCl,
Denatured spirit,	Phosphorous pentoxide,	Senna powder

Procedure

To 200 g of powdered leaves with benzene 600 ml for 2 hr on electric shaker, filter in vacuum and distill off solvent. Dry the marc left after benzene extraction at room temperature and extract with 70 % methanol

600 ml on shaker for 4-6 hr. Filter under vaccum, re extract marc with 400 ml of 70% methanol for 2 hr, filter and combine methanolic extract. Concentrate methanolic extract to one eight volume, acidified to PH 3.2 by addition of HCl with constant stirring. Set aside the mixture for 2 hr at 5 °C, filter under vaccum and to the filtrate add anhydrous $CaCl_2$ (2 gm) in 25 ml of denatured spirit with vigorous stirring. Adjust the PH of the solution to 8 by ammonia solution and set aside for 2 hr. Filter the solution by vacuum and dry precipitate on phospharous pentoxide in dessicator.

1.8   Isolation of Lactose from Milk

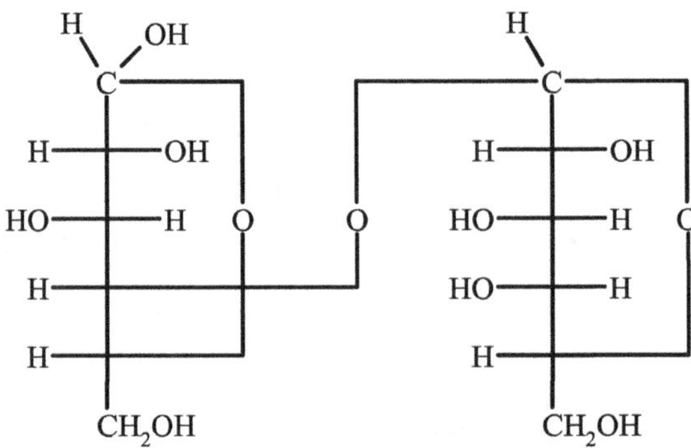

Lactose (milk sugar) can be isolated from milk. It is a disaccharide.

Materials

Milk,	Acetic acid,	Calcium Carbonate,
Ethanol,	Charcoal	

Procedure

Place 200 ml of skimmed (not fat) milk in a 500 ml beaker and warm it to 40 °C then add dropwise few ml of 10% acetic acid until there is a significant precipitate of casein. Do not add excess acid as it may hydrolysis the disaccharides, lactose to the monosaccharide glucose and galactose. Stirr the casein until it forms a large amorphous mass and remove it from the mixture. Add 5 g calcium carbonate to the clear solution. Heat this mixture to a gentle boil for about 10 min.

This should result in the nearly complete precipitation of the albumins. Filter the hot mixture by vacuum to remove the precipitated albumins and excess of calcium carbonate and concentrate the filtrate in a 500 ml beaker, using a water bath to about 30 ml. Then add 175 ml of 95% ethanol and 2 g of decolorizing charcoal to the hot solution. After it has been mixed well, filter the warm solution by vacuum through a layer of wet filter aid. (diatomaceous earth). Transfer the filtrate to a flask and allow it to stand for several days. Lactose crystallizes on the walls and down of the flask. Dislodge the crystals and vacuum filter them. Wash the product with 10 ml of 25% ethanol.

1.9   Color Tests of Carbohydrates

Molisch Test

In this test concentrated H_2SO_4 forms furfurals due to its dehydrating action on the carbohydrates, and the furfurals or intermediate products between the carbohydrates and the furfurals condense with α-naphthol to form colored products.

A negative result by this reaction is very good evidence of the absence of carbohydrates, but a positive test is simply an indicator of the probable presence of carbohydrates.

In each of three test tubes are place 1ml of 0.02 M glucose, 0.01 M Xylose and 0.7% starch. Add 2 drops of fresh 5% naphthol solution in alcohol directly to each substance. Mix each well, then allow 3 ml of concentrate sulfuric acid to flow down the sides of each tube to form a violet ring.

Seliwanoff's

This is an indicator test for hexose.The color developed here is due to the formation of 4-hydroxy methyl-furfural and the reaction of this with resorcinol in the presence of a 12% HCl solution.

Prepare fresh seliwanoff's solution by adding 3.5 ml of 0.5% fresh resorcinol to 12 ml of concentrate HCl and dilute to 35 ml with distilled water. Mix well and measure off 5 ml portions in to test tubes containing respectively, 1 ml of 0.1M glucose, 0.1M Galactose, 0.1M Lavulose. Mix well and immerse the test tubes in boiling water. Observe the color change every 5 min during 15 min boiling.

Phloroglucinol

This reaction is positive for pentoses, but not a specific indication of them, since hexoses and glycoronic acid also react. If the cherry red color develops very promptly and if this is soluble in amyl alcohol and the alcoholic solution, shows a specific absorbance band consistent with resource data, then it can be considered a fairly reliable test for pentoses and glucoronic acid.

Prepare a fresh reagent by mixing 10 ml of concentrate HCl with 2 ml of the phloroglucinol solution and dilute to 18 ml with water. Measure off 5 ml portions of this solution and add, respectively to tubes containing 0.5 ml of 0.2 M Glucose, 0.2 M Xylose solutions. Place all the test tubes in the boiling water and note the color changes over 15 min.

Orcinol

Prepare a fresh reagent by mixing 10 ml of conc. HCl with 2 ml of a 6% aqueous solution of orcinol and dilute it to 18 ml with water. Now measure off 5 ml portions of the orcinol reagent into test tubes containing 0.5 ml of 0.2M Glucose, 0.2M Xylose solutions. Place all the test tubes in the boiling water and note the color changes.

Terpenoids

2.1 Introduction

The term terpene represents only the hydrocarbons $(C_5H_8)_n$ while the term terpenoid represent the hydrocarbons as well as the oxygenated derivatives. In our daily practice we see that fruits, flowers, leaves, stems, barks and roots of nearly all the plants have some pleasant smell. It has been observed that this pleasant smell of the fruits is actually due to the presence of certain steam volatile oils known as essential oils.

Isolation

The isolation of mono and sesquiterpenes is effected in two steps, isolation of a essential oil from the plants and separation of the individual terpenoids from the essential oils.

Isolation of Essential Oils

Plants containing essential oils usually have the greatest concentration at some particular time e.g., jasmine at sunset, hence it is better to take the plant parts having essential oil at this particular time. Three methods have been developed for the isolation of essential oils

- Expression
- Steam distillation
- Extraction

Expression: The plant material is crushed and the juice is screened to remove the large particles. The screened juice is centrifuged in a high speed, when nearly half of the essential oil is extracted, the other half of the oil is generally not extracted and such residue is used for the isolation of interior quality of oil by distillation. Example Citrus, lemon and grass oil are extracted by this method.

Steam distillation: This is the widely used method, the plant material is macerated and then steam distilled when the extracted oil go into distillate from which they are essential by the use of pure organic volatile solvents like light petroleum.

Extraction: If some essential oil are sensitive to heat and hence decomposed during distillation, in such cases the plant material is directly treated with light petrol at 50 °C, and the solvent is then recovered by distillation under reduced pressure.

Separation of Terpeniods

Tilden 1887 discovered that whenever terpeniod hydrocarbons are treated with nitrosyl chloride in chloroform, crystalline adducts having sharp melting point are obtained. The adducts were separated and decomposed in to their corresponding hydrocarbons. The terpenoid alcohols were separated by their reaction with phthalic anhydride to form diesters. The diesters extracted with $NaHCO_3$ and then decomposed by alkali to the parent terpenoid alcohol, terpeniod aldehyde and ketones were separated by reaction with the common carbonyl reagents such as $NaHCO_3,$ 2, 4 di nitro phenylhydrazine, and Grignard reagent.

Important essential oils with their main terpenoid constituents.

Essential Oil	Terpenoid
Turpentine	Pinene
Bergamot	Linalool
Caraway	Carvone,α-limonene
Coriander	Linalool,pinene
Eucalyptus	Cineole
Jasmine	Linalool
Lavender	Linalool
Lemon	d-limonene, citral

Contd…

Essential Oil	Terpenoid
Sweet orange	d-limonene
Peppermint	Menthol
Rose	Geraniol, citronellol
Cardamom	Terpineole
Sandle wood	Santalol
Ginger	zinziberine

Terpenoids are widely distributed in nature, mostly in the plant kingdom. They may be regarded as derivatives of oligomers of isoprene $CH_2 = CH\text{-}CH = CH_2$, usually joined head and tail.

Terpene hydrocarbons are classified as follows

Monoterpenes	$C_{10}H_{16}$
Sesquiterpenes	$C_{15}H_{24}$
Diterpenes	$C_{20}H_{32}$
Triterpenes	$C_{30}H_{48}$
Tetra terpenes or Carotenoids	$C_{40}H_{64}$
Polyterpenes	$(C_5H_8)n$

Monoterpenes

Acyclic Monoterpenes

Those having an open chain structures

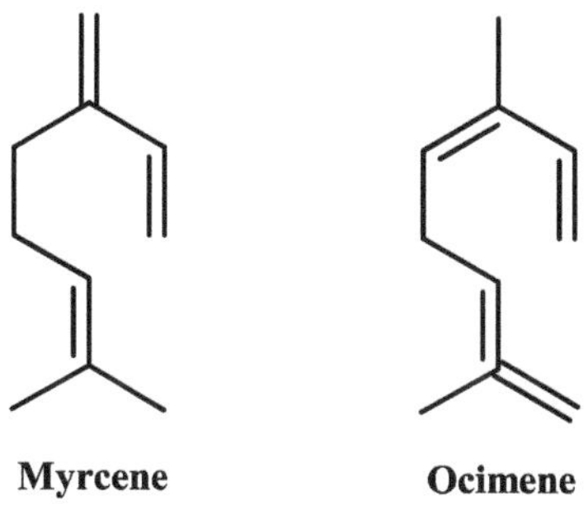

Myrcene   Ocimene

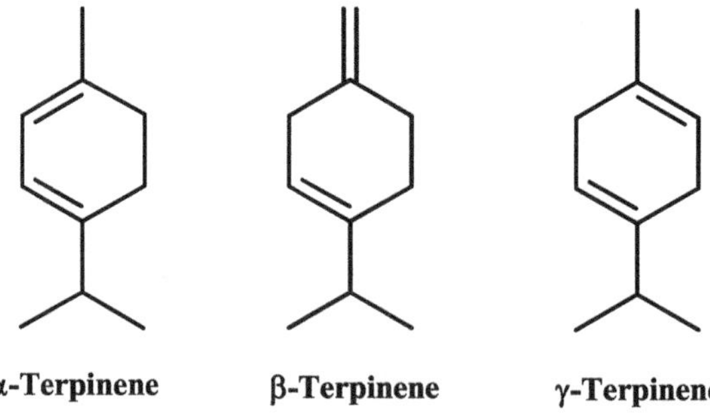

Citronellol

Linalool

Geranial

Citronellal

Geraniol

Citral or Neral

Monocyclic Terpenes

α-Terpinene

β-Terpinene

γ-Terpinene

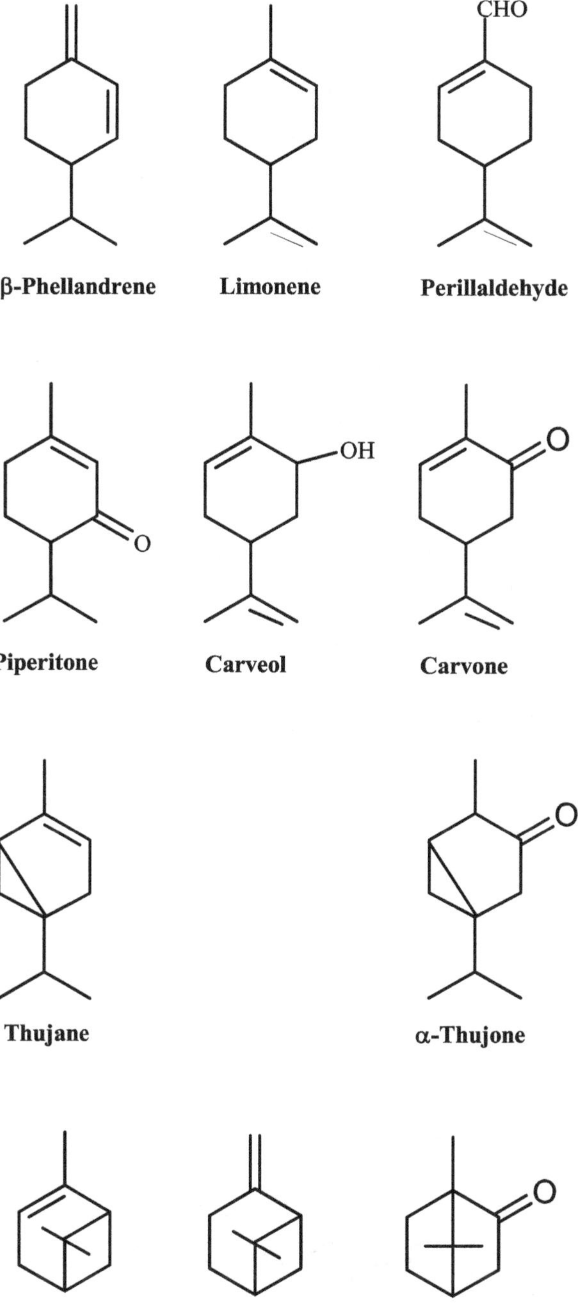

β-Phellandrene   Limonene   Perillaldehyde

Piperitone   Carveol   Carvone

Thujane   α-Thujone

α-Pinene   β-Pinene   Camphor

Sesquiterpenes

The sesquiterpenes form the higher boiling fraction of the essential oil. They are formed by the union of three isoprene units. Sesquiterpenes are unsaturated compounds and may be acyclic monoterpenes, Monocyclic terpenes, Bicyclic terpenes, Tricyclic terpenes, etc.

Diterpenes

The diterpenes are formed by the union of four isoprene units and are found mainly in the resins of plants.

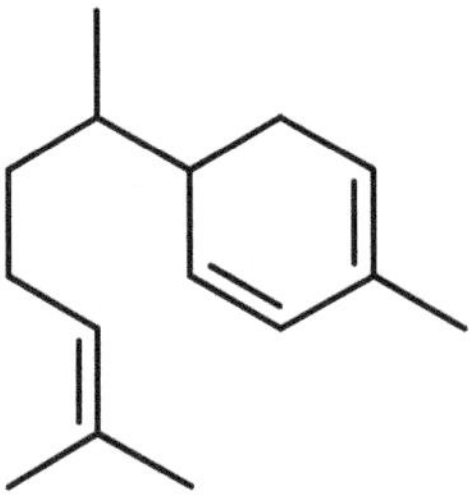

Zingiberene 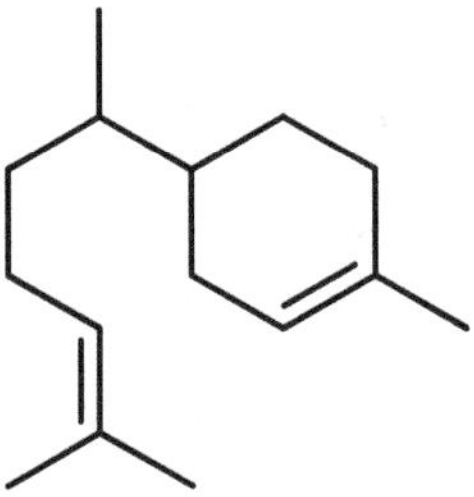   **Bisabolene**

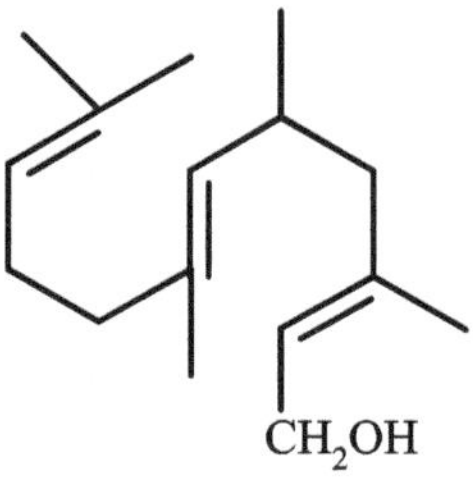

CH_2OH

Farnesol

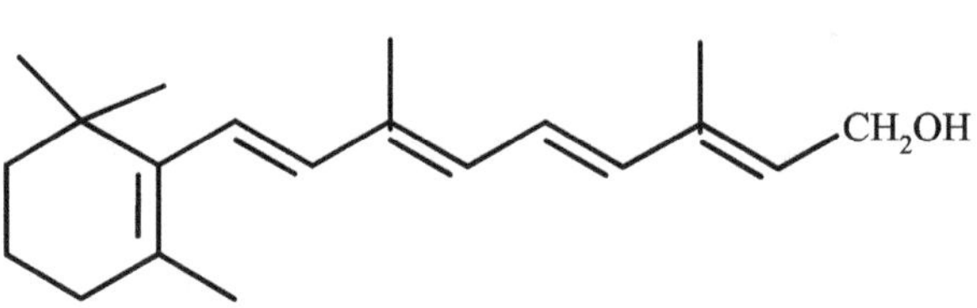

CH_2OH

Vit-A

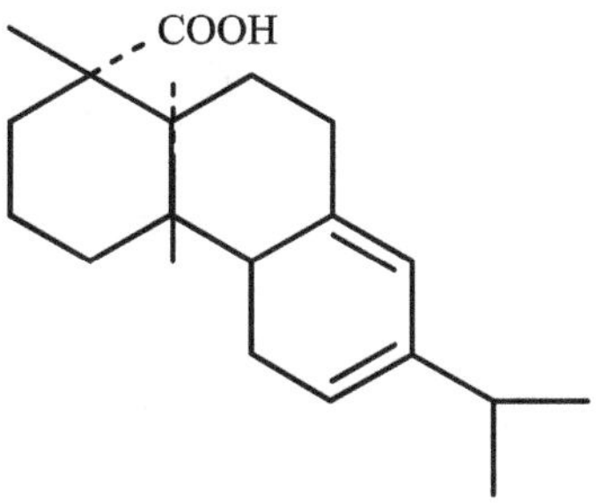

Levopimaric acid

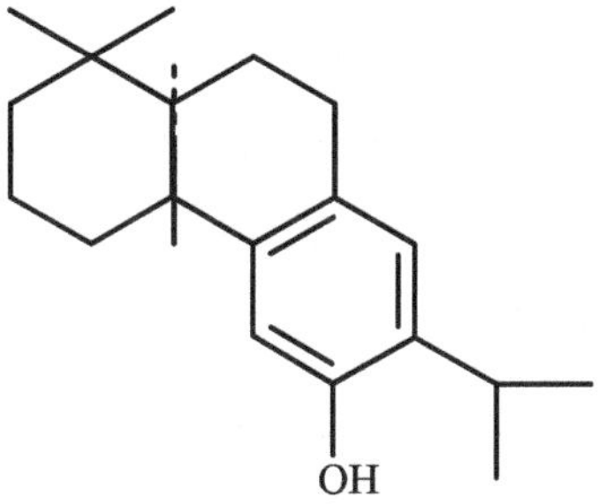

Ferruginol

Triterpenes

The triterpenes are widely distributed in the plant and animal kingdom, where they occur either in the free state, as esters, or as glycosides.

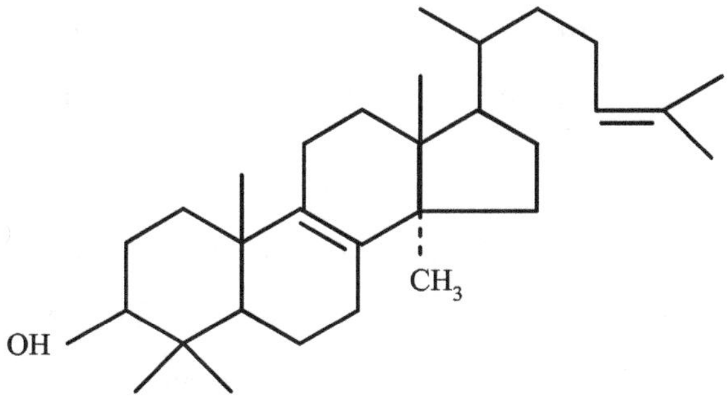

Lanosterol

Squalene

Carotenoids

The carotenoids are yellow or red pigments which are widely distributed in plants and animals. Chlorophyll is always associated with the carotenoids carotene and lutein; the carotenoids acts as photosensitizer in conjunction with chlorophyll. When chlorophyll is absent in fungi then the carotenoids are mainly responsible for color. These are also known as lipochromes chromo lipids because these are fat soluble pigments. In the higher plants the carotenoids are found in the leaves together with chlorophyll, they also constitute the principal pigments of certain yellow, orange, red flowers and many microorganisms.

Chemically carotenoids are polyenes and almost all the carotenoid hydrocarbons have the molecular formula $C_{40}H_{56}$. Since the carbon skelton of these compounds has a polyisoprene structure; they may be regarded as tetra terpenes.

The compound of a long conjugated chain comprised of four isoprene units, the centre two of which are joined tail to tail. The compound may be two open chain structure or one open chain structure and one ring or two rings. The color of the carotenoids is attributed to the extended conjugation of the central chain. X ray analysis has shown that the double bonds in the majority of natural carotenoids are in the trans position and a few natural carotenoids are cis-trans.The carotenoids comprises two groups – Hydrocarbons, soluble in petroleum ether, and xanthophylls, oxygenated derivatives of the carotenes. These compounds are alcohols, aldehyde, ketones, epoxides and acids soluble in ethanol.

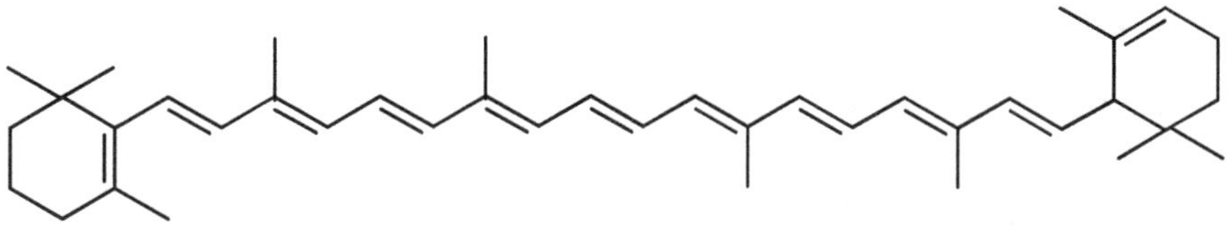

α-Carotene

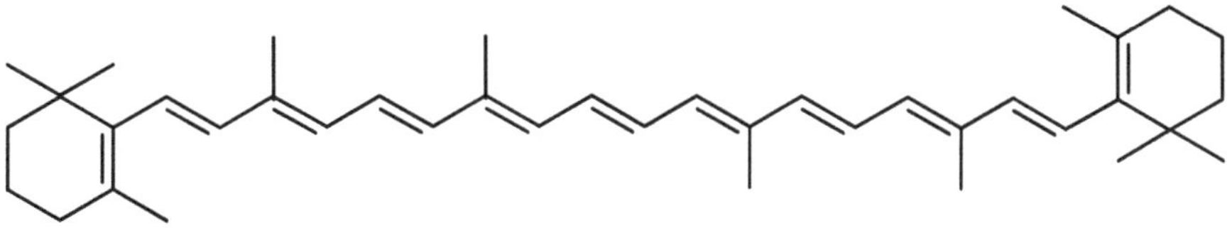

β-Carotene

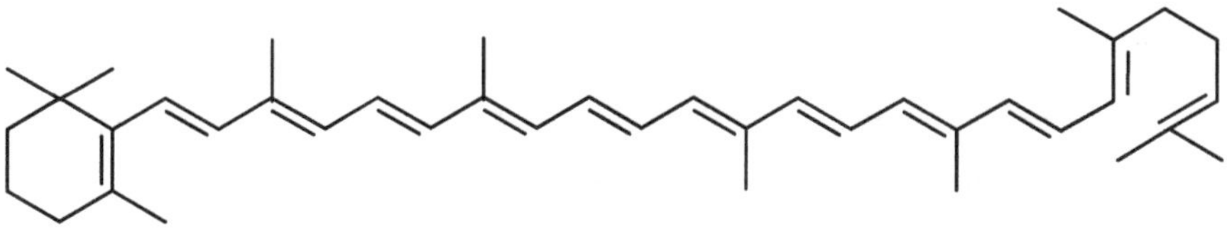

γ-Carotene

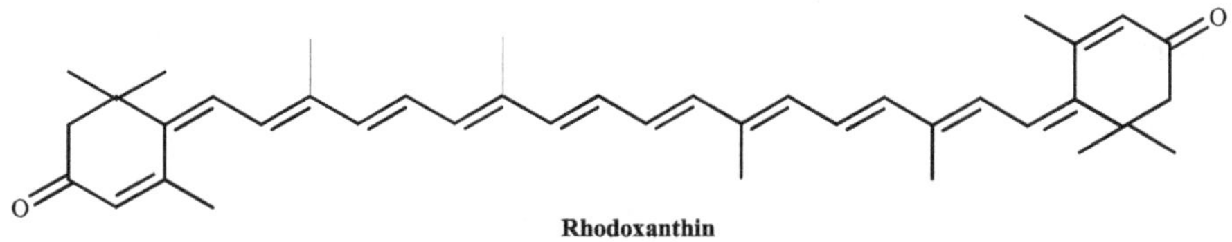

Rhodoxanthin

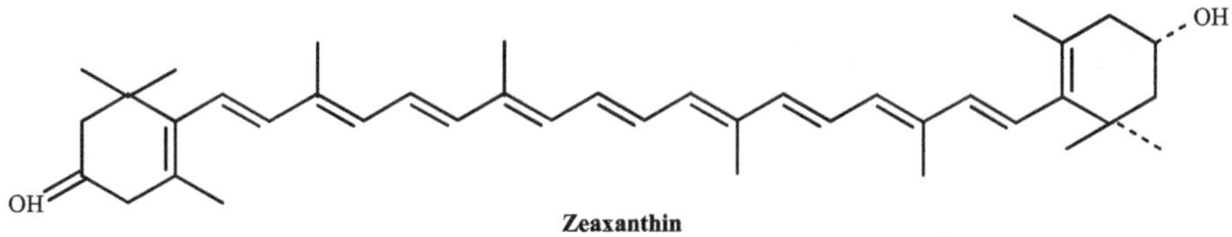

Zeaxanthin

2.2   Isolation of Lycopene from Tomatoes

Introduction

Lycopene was first isolated from *Tamus communis* by Hartsen in 1873. Recent investigation employing highly defined chromatographic methods have shown that the tomato pigment is widely distributed in nature. It has been prepared from a variety of fruits and berries. The first modern preparation is that of Willstatter and Escher, who processed 75 kg tomato concentrate and obtained 11 g once recrystallized lycopene. Recently four dominant tomato paste carotenoids, phytofluene, β-carotene, phytone and lycopene were identified using high performance liquid chromatography with US-VIS photodiode array detection.

Principle

Tomato paste is dehydrated with methanol and lycopene is extracted from the residue with methanol-carbon tetrachloride. The crude product is crystallized twice from benzene by the addition of methanol, giving lycopene of 98 to 99% purity. Further purification is achieved by a chromatographic procedure, using calcium hydroxide as the adsorbent.

Materials

Tomato paste,	Benzene
Carbon tetra chloride,	Methanol
Sodium sulphate (anhydrous)	

Procedure

Canned tomato paste, 50 g in a 3 lit wide mouthed bottle is dehydrated by adding 65 ml methanol. The mixture is immediately shaken vigorously to prevent the formation of hard lumps. A small sample of the suspension is tested by hand; if it has glutinous consistency, more methanol is added to the main portion to avoid the possible clogging of filters.

The mixture is allowed to stand for 1 to 2 hr and is then shaken vigorously. The thick suspension is filtered on a Buchner funnel. The yellow filtrate is discarded. The dark red cake is returned to the bottle and shaken with a mixture of 35 ml methanol and 35 ml CCl_4. The stopper of the bottle must fit well and should be lifted for a moment after the mixing to release any built up pressure. Brief shaking followed by opening of the bottle is repeated until no more excess pressure is noticed.

The suspension is shaken for 10-15 min and separated by filtration on a large Buchner funnel. The filtrate consists of a lower, very dark red CCl_4 phase and an orange aqueous-methanolic layer. The slightly tomato colored residue is crushed by hand to form a nearly uniform powder. It is then reextracted with 35 ml of each solvent as described, and the suspension is filtered. The extraction is now almost complete.

The filtrates are combined, the methanol layer is transferred to a 2 lit seperatory funnel and 1 volume of water is added. A white emulsion appears in the upper phase. If the emulsion is reddish, it is stirred with a glass rod until the droplets of CCl_4 join the lower layer. The phases are separated and the CCl_4 phase is washed several times with water. The CCl_4 solution is drained into a 1 lit Erlenmeyer flask and dried over anhydrous sodium sulfate.

The extract is then poured through a folder filter into a 1 lit round bottomed flask equipped with a standard taper. The solvent is evaporated with the aid of a water pump to about 5 ml in a water bath at 60 °C. The solution is transferred to a similar flask of 10 ml capacity using a few ml CCl_4 to rinse the larger flask.

The solvent is then removed completely *in vacuo*, leaving a dark oily residue, which is diluted with a few ml benzene and evaporated again to remove the CCl_4 completely. The partly crystalline dark residue is transferred quantitatively with 1 ml benzene to a 25 ml Erlenmeyer flask. A clear solution is obtained by immersing the flask in a hot water bath. Boiling methanol is added in portions, using a dropper, to the benzene solution, with stirring after each addition, until 1 ml methanol has been introduced. Crystals of crude lycopene begin to appear immediately.

The crystallization is completed by keeping the liquid first at room temp. and then in ice water. After standing for 1 to 2 hr, the crystals are collected on a small Buchner funnel and washed with 2 ml boiling methanol. The lycopene crystals are transferred to a tared 10 ml centrifuge tube, the last portion being removed from the funnel with small quantities of boiling benzene. Benzene is added to the centrifuge tube to make up the volume to 1 ml.The crystals are dissolved by dipping the tube into hot water and stirring the contents.

When a clear solution is obtained boiling methanol is introduced in small portions with a dropper and the solution is stirred with a glass rod until crystals begin to appear. The centrifuge tube is kept at room temperature for a short time and then an ice bath. More methanol is added in small portions, with stirring, to the cold solutions. The total volume of methanol present should not exceed ice bath and the crystals are separated by a brief but strong centrifuging. The mother liquor is decanted and discarded. The crystals are treated in the centrifuge tube with 1 ml boiling methanol.

The mixture is stirred and the methanol is removed by centrifuging before it cools. The methanol is decanted, and purification was satisfactory, long, red lycopene prisms are observed under the microscope. No color substance should be present. The centrifuge tube and its contents are dried in vacuo at room temp for a few hr and then weighed. The yield, which is dependent on the quality of the tomato paste, is about 1.5 mg.

Chemical Test

1. Lycopene dissolves in concentrate Sulfuric acid imparts indigo blue color

2. On adding a solution of antimony chloride in chloroform to a solution of lycopene in chloroform an intense unstable blue color is produced.

2.3   Isolation of Lemon Grass Oil from Lemon Grass

Introduction

Lemon grass oil is a volatile oil or essential oil contains chemical constituents like limonene [terpene] and citral [terpenoid] etc. Lemon grass oil is pharmaceutically used as flavoring agent and carminative. Volatile oil is immiscible with water that can be extracted by simple distillation by using water due to slight solubility at boiling temperature. The important constituent of lemon grass oil is citral (trans form is called CITROL and cis form is called NEROL) which is categorized under acyclic mono terpenoid that exhibit aldehyde as functional group. It can be estimated by oxime formation principle based up on the aldehyde group (citral content in lemon grass oil is 3.4-3.6%). The estimation of the citral in isolated oil can be estimated by oxime formation in which the citral is treated with hydroxylamine HCl as a result, the formation of citral oxime with equimolar liberation of HCl. The liberated HCl is back titrated with KOH using methyl orange as an indicator.

However, the titration method of estimation is limited due to the low active constituent concentration and the alternative method of estimation i.e., quantitative gas chromatography method that especially allows the sample of volatile nature. The GC method allows the samples to separate up on the solubility of volatile constituents between carrier gas and stationery phase. GC method may be applied to estimate the volatile chemical constituents in the sample as well as the identification of the same by comparing to the standard.

Materials

Lemon grass leaves

Distill water

Procedure

To the slurry of lemon grass add distill water in excess and transfer in to a round bottom flask. Keep Round bottomed flask in distillating unit heat at less than 100°C provided that the temperature was not less than the aqueous phase boiling point. The distillate was collected and allowed to stand a side for 20 min. The layer of volatile oil that is immiscible with water phase was separated by capillary method and subjected to estimation of aldehyde content and identification.

2.4   Isolation of β-carotene from Carrots

Introduction

The carotenoids are the yellow (or) red colored pigments widely associated with chlorophyll as photosensitizer. Carotenoids are also known as lipochromes or chromo lipids. Due its lipid solubility some carotenoids are hydrocarbons (carotene) and some are oxygenated derivatives which are called xanthophylls. Chemically the carotenoids are polymers (poly unsaturated) having molecular formulae of $C_{40}H_{56}$ regarded as tetra terpenoids in which the isoprene units are linked by tail-tail pattern at the centre of the molecule (exception for special isoprene rule, that states that isoprene units are linked in terpenoids and terpene with tail-head pattern). Carotenoids exhibits geometrical isomerism as transform.

Materials

Carrot strips

Carbon Tetra Chloride (CCl$_4$)

Sodium sulphate (anhydrous)

Benzene

Methanol

Procedure

Weigh 500gm of carrot and cut in to thin strips and then leave it overnight or for a day or two so as to remove water from it. Powder the dried carrot strips in a pastel mortar and then transfer it in a 2 lit round bottom flask fitted with a condenser. Add 1200 ml of carbon tetra chloride and shake the flask vigorously. Heat the flask on boiling water bath for 3-4 min. Filter the yellow filtrate and store separately. Repeat the extraction process twice each time using 1200 ml of CCl_4. Combine the three organic layers and pour in a separating funnel containing 1000 ml water. Shake the flask thoroughly, remove the lower carbon tetra chloride layer and dry over anhydrous sodium sulphate. Filter the solution, transfer it in a 5 lit round bottom flask and distill off the solvent on steam bath. Transfer the dark oily residue to porcelain dish, add 10 ml of benzene to it and evaporate it off on water bath to remove CCl_4 completely.

To this partly crystalline residue add 25 ml of benzene, heat on water bath to get a clear solution. Cool, add drop wise 5 ml methanol and again cool in an ice bath. Collect the crystals of β-carotene and dry, m.p 183 °C.

Carr Price Test

To the Carotenoids add H_2SO_4 and chloroform solution of anti-mercuric chloride gives blue color which is used in the quantitative estimation of β-carotene and vitamin-A.

2.5   Isolation of Eugenol from Cinnamon Leaf

Introduction

Cinnamon leaf oil is obtained from the leaves of cinnamomum zeylanicum family *Lauraceae*. It is good source of eugenol, the main constituents of the oil were eugenol (76.60%), linalool (8.5%) and piperitone (3.31%). The composition of the oil is comparable to cinnamon leaf oil produced in Bangalore and Hyderabad (south India) in terms of eugenol and linalool contents. Eugenol exhibits antioxidant and anti-inflammatory activities, but at higher concentrations acts as an oxidant and potent allergen. It was earlier shown that *bis*-eugenol synthesized by the oxidation of eugenol was less cytotoxic and more highly antioxidative than eugenol.

Materials

Solvent ether

Potassium hydroxide

Cinnamon leaf oil

Procedure

Dissolve 10 ml of volatile oil in 100 ml of solvent ether and shake with 3 successive quantities of 100 ml of percent potassium hydroxide solution. Regenerate eugenol by acidifying the aqueous layer with excess of sulphuric acid and extracting the acid layer with 3 successive quantities of 50 ml of ether. Distill off the solvent with care to ensure minimum loss of eugenol and dry the residue in desiccator.

Uses

Flavoring agent in a variety of foods and pharmaceutical products.

Mild rubefacient in dentifrices

Chemical test

To 5 ml of compound add 2-3 drops of neutral $FeCl_3$ solution. A yellow green colored solution is obtained

To 5 ml of compound in 1 ml chloroform add 1 ml picric acid solution in chloroform. Heat the solution on water bath for few min and cool. Brown red color crystals of picrate are obtained.

Spectral data

UV spectroscopy

Solution	λmax (nm)
0.1N NaoH	296
Ethanol 95%	281 and 230

IR spectroscopy (cm^{-1})

O-H	3453
C-H	3009
C = C-H	2980
C =C (alkene)	1600
C = C (aromatic)	1520

^{1}HNMR (δppm):

Solvent: $CDCl_3$

($-CH_2$, d)	3.26 - 3.34
($-OCH_3$, s)	3.85
($=CH_2$, m)	4.9 - 5.5

6.6 – 7.2 (aromatic protons, m)

s = singlet   d = doublet   m = multiplet

^{13}C NMR (δppm):

Solvent: $CDCl_3$

Chemical shift	**Carbon No.**
143.8	C_1
146.8	C_2
114.2	C_3
131.8	C_4
121.1	C_5
115.4	C_6
39.8	C_7
137.7	C_8
111.1	C_9
55.87	$(-OCH_3)$

EI – Mass (m/z, %):

164 (base peak, 100)

149 (22)

137 (18)

133 (20)

91 (15)

77 (14)

3

Alkaloids

3.1 Introduction

Konigs 1880 suggested that alkaloids should be defined as naturally occurring organic bases which contain pyridine ring. This definition, however embraces only a limited number of compounds, and so the definition was again modified a little later by Ladenburg, who proposed to define:

"Alkaloid is a natural plant compounds having a basic character and containing at least one nitrogen atom in a heterocyclic ring"

Alkaloids are usually found in the seeds, roots, leaves or bark of the plant and generally occur as salts of various plant acids are isolated by extraction with dilute acids (HCl, H_2SO_4, acetic acid) or alcohol. If these are present as salts, they are liberated by treatment with calcium hydroxide before extraction. The individual alkaloids are usually separated from the very complex mixtures in which they occur by chromatographic methods. The presence of alkaloids is detected by either precipitants or color reagents. The more important precipitants are those of Mayer, Wagner, Dragondroff and Bertrand etc. The color reagents mostly consist of dehydrating or oxidizing agents or a combination of the two.

General Method of Isolation

The presence of an alkaloid is ascertained in the experimental plant for which the plant extract is treated with various alkaloidal reagents such as tannic acid, picric acid, picolinic acid, perchloric acid, potassium mercuric iodide, Dragendorff's reagent, phosphomolybdic acid and phosphotungstic acid, with which the alkaloids either give a precipitate or turbidity.

The dried powdered plant material is first extracted with petroleum ether and then filtered for the removal of soluble fats. The residue is then extracted with methyl alcohol to remove cellulosic and other insoluble materials and the filtrate so obtained is evoperated. The evaporated mass is dissolved in water, acidified to PH 2 and finally steam distilled to remove methyl alcohol. The dark residual solution is either allowed to stand for several days in a refrigerator or heated with molten paraffin to remove suspended impurities. The filtrate is extracted with ether or chloroform to remove water soluble non basic organic material and then steam distillate, and then the steam volatile alkaloids are separated. The solution of the rest of the alkaloid's salts is made alkaline and again extracted with ether or chloroform and ethereal layer obtained after this extraction is evaporated to give crude alkaloids.

The alkaloids are generally classified by their molecular structure.

- Indole alkaloids
- Quinoline alkaloids
- Isoquinoline alkaloids
- Phenanthrene alkaloids
- Pyrrrolidine alkaloids
- Pyridine and piperdine alkaloids
- Pyrrolidine and pyridine alkaloids
- Tropane alkaloids
- Imidazole alkaloids
- Steroidal alkaloids

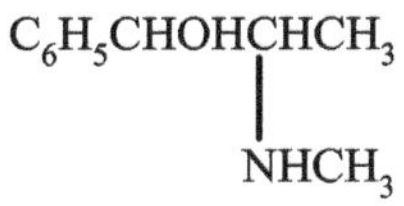

$$C_6H_5CHOHCHCH_3$$

with $NHCH_3$ substituent

Ephedrine

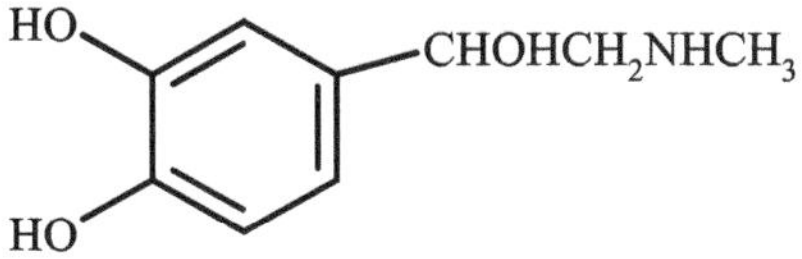

Adrenaline

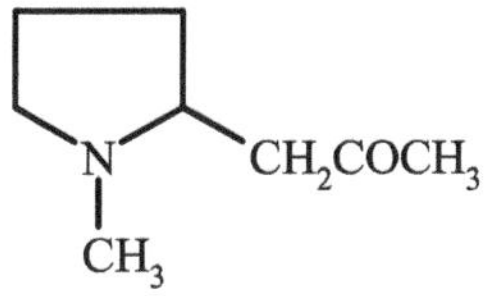

Hygrine

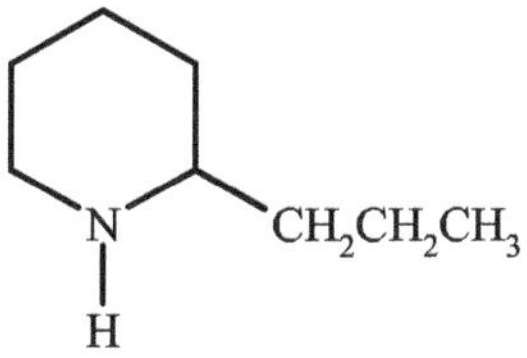

Coniine

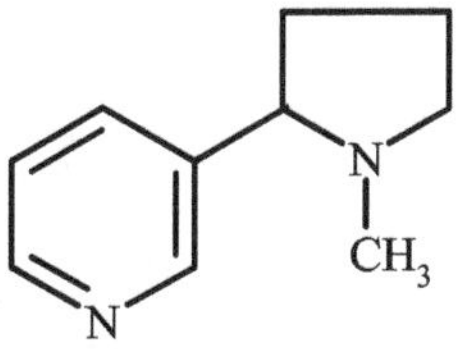

Nicotine

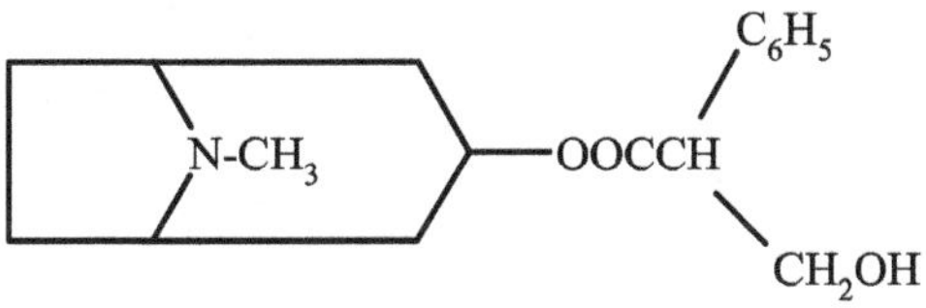

Atropine

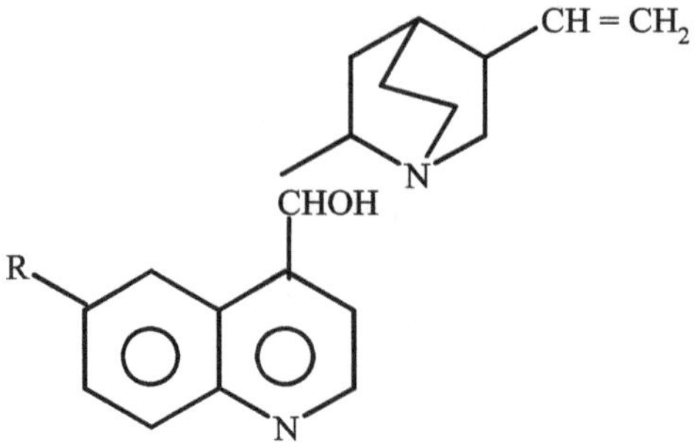

Quinine (R = OCH$_3$)
Cinchonine (R = H)

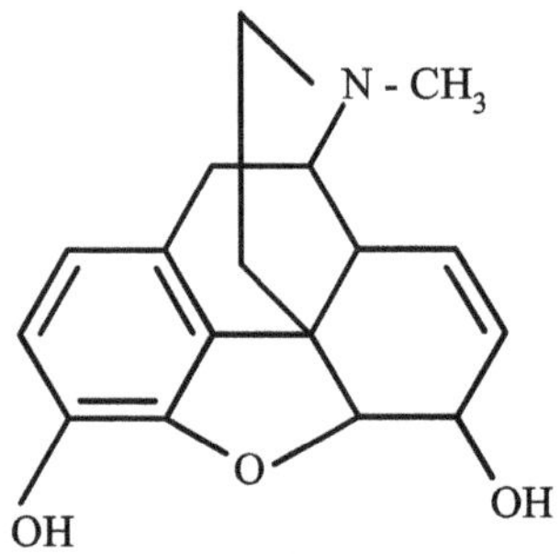

Morphine

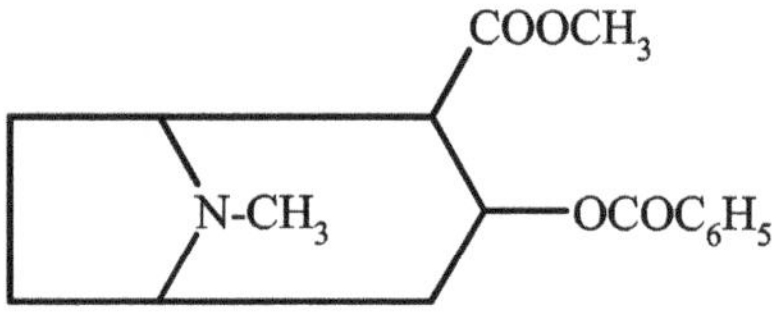

Cocaine

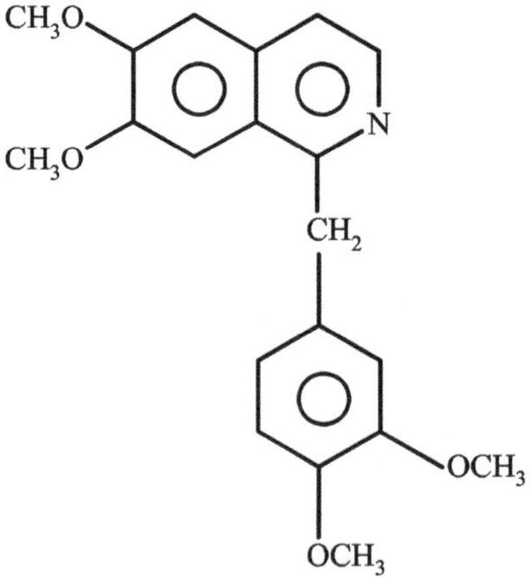

Papavarine

Reserpine

3.2   Isolation of Caffeine from Tea Dust

Introduction

Caffeine is one of the most important naturally occurring methyl derivatives of xanthine. Its concentration in a variety of teas, including black and green teas, depends both on climatic and topographic conditions of growth and on processing methods and was found to vary from 2.0 to 4.6%. Thus Chinese black tea contains 2.6-3.6%, brazilian 2.2-2.9% and Turkish 2.1-4.6% caffeine. In addition to caffeine, other purins have also been reported to be present in small quantities. Michl and Haberler separated the following compounds from black tea: caffeine 2.5%, theobromine 0.17%, theophylline 0.013% and traces of guanine, xanthine and hypoxanthine.

Caffeine is chemically called as 1, 3, 7-tri methyl xanthine, naturally distributed in tea about 2-4% and coffee about 1-2%. Caffeine is xanthine alkaloid. The principle involves the isolation of caffeine from tea leaves by extraction with hot sodium carbonate solution and neutralization and extraction with dichloromethane.

Materials

Tea,	Sulfuric acid,	Celite
Dichloromethane,	Sodium carbonate	

Procedure

Weighed accurately 20 gm of finely powdered tea leaves are placed in a 400 ml beaker, 5 gm sodium carbonate and 100 ml of water is added, and the mixture is heated by a Bunsen burner for 20 min. Water is occasionally added to keep the volume of the solution constant. The hot solution is then filtered, and the filtrate is neutralized cautiously, with stirring, with sulfuric acid 10% solution.

It is then filtered through a thin layer of a filter aid (Celite) placed on a Buchner funnel padded with a wet filter paper, and washed with 20 ml dichloromethane. The two phase filtrate is then placed in a separating funnel. The organic layer is separated, and the aqueous layer is extracted twice with two 40 ml portions of dichloromethane. The two organic layers are then combined, and the solvent is evaporated. Crude caffeine is crystallized from a very small quantity of hot acetone or water. The pure crystalline needles of caffeine melt at 235 °C.

Chemical Tests (Murexide Test)

A few crystals of caffeine and few drops of nitric acid are placed in a small porcelain dish and evaporated to dryness followed by addition of two drops of ammonium hydroxide imparts a purple coloration.

Spectral Data:

UV spectroscopy:

Solvent	λmax	E1%, 1cm
Methanol	272	-
Trichloroethylene	278	-
Ethanol	273	519
0.1N HCl	272	470
0.1N NaOH	275	490

IR spectroscopy (KBr disc, cm^{-1}):

3110	-CH$_3$str
2950	-CH str
1700	C = O
1650	C = N

PNMR spectroscopy:

Solvent: CDCl$_3$

TMS as internal reference

Chemical shift (ppm)	**Group**
3.53 (s)	1-N-CH$_3$
3.33 (s)	3-N-CH$_3$
3.98 (s)	7-N-CH$_3$
7.54 (s)	8-H

S = Singlet.

^{13}C NMR spectroscopy:

Carbon No.	**ppm**
C-2	151.69
C-4	148.73
C-5	107.55
C-6	155.35
C-8	141.53
C-10 - CH$_3$	29.70
C-11 - CH$_3$	27.87
C-12 - CH$_3$	33.54

EI – Mass spectroscopy (m/z,%):

Molecular formula: $C_8H_{10}N_4O_2$

Molecular weight: 194

194 (M+, 100)

109 (66)

82 (37)

67 (54)

55 (80)

3.3   Isolation of Piperine from Black Pepper

Introduction

Piperine occurs in unripe fruit (black pepper) and the kernel of the ripe fruit white pepper) of Piper nigrum and in the fruit of aschanti (piper clusii). It is also found in long pepper (piper longum) and in the seeds of Cubeda censii. Piperine has also been isolated from piper famechoni. The piperine content of black pepper varies from 6 to 9%. Piperine is tasteless, so some investigators have suggested that the stereoisomeric chavicine, which accompanies piperine in the pepper, is possibly responsible for the particular taste of pepper. The piperine is extracted from black pepper with ethanol and may be converted to a red TNB (1, 3, 5 tri nitro benzene).

Piperine is tasteless at first and it does ultimately produce a burning sensation and sharp after taste. The initial tastelessness of piperine may be a consequence of its extremely low solubility in water. The crude ethanol extract contains, in addition to piperine and chavicine, some acidic, resinous material. In order to prevent co-precipitation of piperine and the resin acids, dilute ethanol KOH solution is added to the concentrated extract to keep acidic material in solution and /or solid gummy material that is precipitated in the vessel.

Materials

Black Pepper,	KOH,
Ethanol,	Tri Nitro Benzene (TNB)

Procedure

Grind 25 g fresh black pepper corns to a fine powder, place in a Soxhlet thimble, and extract with 100 ml ethanol for 90 minutes. Cool the resulting solution, filter if necessary, and concentrate on the rotary evaporator. Keep the water bath below 60 °C during the concentration. Dissolve the residue in 10 % alcoholic potassium hydroxide (25 ml). Decant the solution if any residue remains. Cool the solution in an ice bath, and add water dropwise (about 30 ml will be required) to precipitate the piperine. Collect the piperine on a sintered glass funnel and dry it on the vacuum pump. If time allows, a recrystallization from Acetone: Hexanes.

MP: 125-126 °C.

IR spectroscopy:

Aromatic C-H (str)	3000
C=C (Diene)	1635, 1608
C=C (benzene)	1580, 1495
CO-N (str)	1635
-CH$_2$ (bend)	1450
Asymmetric C-O-C (str)	1250, 1190
Symmetric C-O-C (str)	1030

EI-Mass (m/z, %):

Molecular formula: $C_{17}H_{19}NO_3$

Molecular weight: 285

285 (M+,100)

3.4   Isolation of Nicotine from Tobacco

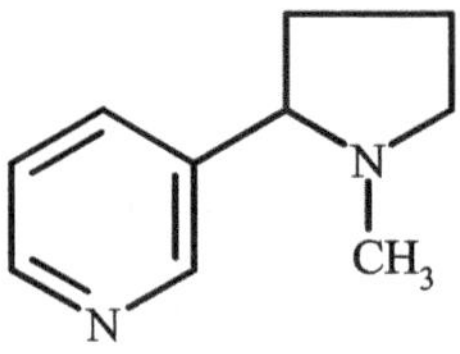

Introduction

Nicotine was first observed in 1809 by a French Scientist, Vauquelin. In 1828, Poselt and Reimann from Heidelberg University, were the first to isolate and purify nicotine. In 1895, Adolf Pinner revealed and acquired the structure of nicotine known today. The first studies of the metabolism of nicotine were in the 1950s.

Nicotine is an alkaloid and it is a highly toxic drug, with only 60 mg being lethal to an adult. The average cigarette contains 8 to 9 mg of nicotine; so one pack of cigarettes contains enough nicotine to kill your average adult, to say nothing of a child. Depending on technique, a smoker receives about 1 mg of nicotine per cigarette.

Nicotine has been used as a stimulant and relaxant for thousands of years. It has been the subject of wars, of peace, of any number of different penalties, and still is widely used today. Nicotine is also a potent natural insecticide. Many millions of pounds were produced and used in the 1940s and 1950s before extensive synthetic pesticides were developed.

Materials

Electric oven,	Refrigerator,	Alc KOH
Dil. H_2SO_4,	Dil. NaOH,	Solvent ether
Picric acid,	Tobacco	

Procedure

Moisten about 20 g of dry powder with sufficient quantity of 20% Alc KOH to liberate alkaloids base. Dry it in oven below 60°C. Place the powder in 250 ml conical flask and add 50 ml of solvent ether, put a cotton plug, heat on electric water bath till it boils, shake for 2 min, boil again and repeat this process for 5 min. Filter and concentrate the filtrate to 20 ml and treat ethanol extract with 25 ml of 1% H_2SO_4 twice.

To aqueous layer, add sufficient quantity of NaOH (5%) solution. Extract free base with 20 ml of solvent ether twice. Concentrate ethorial extract and dissolve residue in about 3-5 ml of distilled water, filter, and to the filtrate, add saturated solution of picric acid drop wise till complete precipitation of nicotine takes place. Keep the solution in refrigerator for half an hour. Remove supernent liquid by decantation, dry the product.

3.5   Isolation of Solanine from Potatoes

Introduction

Potatoes contain glycoalkaloids, toxic compounds, of which the most prevalent are solanine and chaconine. Glycoalkaloids occur in the greatest concentrations just underneath the skin of the tuber and they increase with age and exposure to light. Glycoalkaloids may cause headache, diarrhea, cramps and in severe cases coma and death. However, poisoning from potatoes occurs very rarely.

Solanine is a glycoalkaloid poison found in species of the nightshade family, such as potatoes. It can occur naturally in any part of the plant, including the leaves, fruit, and tubers. It is very toxic even in small quantities. Solanine has both fungicidal and pesticidal properties and it is one of the plant's natural defenses. Solanine occurs naturally in all nightshades, including tomatoes, capsicum, tobacco and eggplant, as well as plants from other species. However, most ingested solanine is from the consumption of potatoes

Materials

Potatoes,	Acetic acid,
Ammonium hydroxide,	Methanol

Procedure

Macerate the pieces of potato tubers with twenty parts of 5% acetic acid for 24 hr. Warm to 70 °C and add conc. ammonium hydroxide drop wise and adjust the pH to 10. Centrifuge, wash precipitate thrice with sufficient quantity of 1 per cent ammonium hydroxide and recentrifuge. Discard the supernent liquid and wash, dry the precipitate and purity it by dissolving in boiling methanol, filtering and concentrating until the glycoalkaloid starts crystallizing out.

3.6   Color Reactions for Alkaloids

Some of the major alkaloidal test solutions are

- **Dragendorff's reagent**   (KI and Bismuth Iodide)

- **Wagner's   reagent**   (2 g of Iodine and 6 g of KI in 100 ml of water)

- **Mayer's reagent**   (1.358 g of $HgCl_2$ and 5g of KI in 100 ml of water)

- **Erdmann's reagent**   (10 drops of HNO_3 in 20 ml of conc. H_2SO_4)

- **Froehd's reagent** (1 g of NH_4MoO_4 in 100 ml of H_2SO_4)

- **Marme reagent** (2 g of CdI_2 and 4 g of KI in 12 ml of water)

- **Sonnenschein's reagent** (Phosphomolybdic acid)

- **Ferric chloride solution** (82 g of ferric chloride in 100 ml of water)

- **Mandolin reagent** (1 g of NH_4VO_4 in 200 g of H_2SO_4)

- **Tannic acid solution** (1 g of Tannic acid in 1 ml methanol + 10 ml water)

- **Picric acid solution** (1 g of picric acid in 100 ml of water)

Reagent	Color	Alkaloid
Dragendorff's reagent	Reddish brown	Piperine and nicotine
PtCl$_6$ solution	Buff colored spot	Tropane
Bromine and ammonia	Emrald green	Quinoline
FeCl$_3$ solution and 5 ml H$_2$SO$_4$ Addition of HNO$_3$	Blue Dark red brown	Isoquinolines
Mandelin	Violet to red Yellow to green finally blue	Indole Imidazole
Wagner's	Reddish brown	Alkaloids
Mayer's	Cream precipitate	Alkaloid
Froehde's	Bluish green	Ipacaunha alkaloids

4

Proteins

4.1 Introduction

The name protein was introduced by Mulder 1839, who derived it from the greek word proteins. Proteins are nitrogen substances which occur in the protoplasm of all animals and plant cells.

The proteins can be broken down into smaller and smaller fragments until the final products as the amino acids

$$\text{Protein} \rightarrow \text{polypeptides} \rightarrow \text{Peptides} \rightarrow \text{Amino acids}$$

Several arbitrary classifications of the proteins are in use. One method divides the proteins in to two groups.

- **Simple proteins**

  Albumins are soluble in water, acids, alkali and are coagulated by heat, include serum albumins and egg albumins.

  Globulins are insoluble in water, but are soluble in dilute solutions of inorganic acids and alkalies; include serum globulin and vegetable globulin.

  Prolamines are insoluble in water but soluble in 70-90% alcohol; include zein from maize, Gliadin from wheat and Herdein from barley.

  Glutamines are insoluble in water, but are soluble in dil. solution of acids and alkalis; they are coagulated by heat, include glutenin from wheat and oryvzenin from rice.

Scleroproteins are insoluble in water or salt solution, but are soluble in conc. acids or alkalis; include keratin from hair, fibroin from silk.

Histones are soluble in water or dilute acids, but are insoluble in dil. ammonia. They are not coagulated by heat and contains large amount of histidine and arginine.

Protamines are more basic than the histones and have a simpler structure. They are insoluble in water, dil. acids and dil. ammonia. They are not coagulated by heat, and are precipitated from solution by ethanol; include trypsin and papain.

- **Conjugated proteins**

Nucleo proteins on hydrolysis yield protein and prosthetic group. Nucleic acid are present in granular tissue and yeast.

Chromo proteins are characterized by the presence of a colored prosthetic group, include chlorophyll and hemoglobin.

Phospho proteins are conjugated protein in which the prosthetic group contains phosphoric acid.

Glyco proteins are contains prosthetic group a carbohydrate or a derivative of the carbohydrate.

- **Derived proteins**

These are degradation products obtained by the actions of acids, alkalies or enzymes on proteins.

Denatured proteins are insoluble protein formed by the actions of heat.

Meta proteins are insoluble in water or dilute salt solution, but are soluble in acids or alkalis.

Proteases are soluble in water, not coagulated by heat and are precipitated by saturation with ammonium sulphate.

Peptones are soluble in water, not coagulated by heat and are not precipitated by saturation with ammonium sulphate.

4.2   Isolation of Casein from Milk

Introduction

Casein (from Latin caseus "cheese") is the predominant phosphoprotein that accounts for nearly 80% of proteins in milk and cheese. Milk-clotting proteases act on the soluble portion of the caseins. Casein consists of a fairly high number of proline peptides, which do not interact. There are also no disulfide bridges. As a result, it has relatively little secondary structure or tertiary structure. Because of this, it cannot denature. It is relatively hydrophobic, making it poorly soluble in water. It is found in milk as a suspension of particles called casein micelles which show some resemblance with surfactant-type micellae in a sense that the hydrophilic parts reside at the surface.

The caseins in the micelles are held together by calcium ions and hydrophobic interactions. Casein is phosphoprotein and also principle protein found in milk at 35 gm/lt conc. Apart from casein, milk serves as source for lactose, which is soluble in water and contain phospodin, which causes sedation in children. The isolation of casein from milk is based upon the isoelectric point of casein which fall at PH between 4.6-4.8 hence the milk is acidifying with glacial acetic acid to isoelectric point there by casein is precipitated and isolated. The purification of pure casein is obtained by the solubility of principle casein in which isolated casein is treated with ethanol, it removes unwanted fatty material and hence casein remain insoluble in ethanol. Precipitation may also aided by sodium acetate buffer

Procedure

100 ml of milk was taken into 500 ml beaker and warmed at 40 °C. The glacial acetic acid was added drop wise with continuous stirring and adjusted the pH 4.6. The precipitated casein was separated and dispersed in ethanol to solubulise the unwanted fatty material, filtered and treated against ether to remove fractions of alcohol and insoluble fatty content. It was dried at 40 °C and stored in air tight container. The determined and the yield was reported in gm/lit.

4.3   Isolation of Lysozyme from Albumin

Introduction

Lysozyme is one of the smallest protein. It has the ability to break down the cells of certain bacteria. Lysozyme has been obtained from albumin as the insoluble flavianate by procedures involving solvent fractionation and precipitation. More recently it was isolated by adsorption on bentonite and elution with 5% pyridine solution adjusted to pH 5.3, by direct crystallization at pH 9-9.5, by use of an ion exchange resin and by chromatography on hydroxyl apatite (calcium phosphate adsorbent). Lysozyme has a mol wt. of approximately 17,500 and is a basic protein with an isoelectric point of 10.5 to 11.0. It is regarded as an antibiotic since it is able to kill microorganisms and even to dissolve living bacterial bodies. Lysozyme is found in various animal species and is also produced by bacteria. E.g., bacillus subtilis and staphylococcus aureus.

Lysozyme is adsorbed on bentonite together with other albumin globulins. The contaminating proteins are removed from the bentonite by washing successively with alkaline phosphate and 5% aqueous pyridine. The lysozyme is then eluted from the bentonite with 5% aqueous pyridine adjusted to pH 0.5 with sulfuric acid. The eluate is dialyzed, Amorphous lysozyme is obtained by lyophilization of the solution.

Materials

Bentonite,	Eggs,	Pyridine
Sodium chloride,	Sodium hydrogen carbonate	
Phosphate buffer,	Potassium chloride	

Procedure

To 300 ml homogenized albumen are added 50 ml 10% bentonite suspension in 1% potassium chloride. The mixture is stirred vigorously for

3 to 5 min until smooth suspension is obtained. The clay is separated by centrifugation and washed twice with 100 ml portions of 0.5M phosphate buffer (pH 7.5) and three times with 150 ml portions of 5% aqueous pyridine. The clay is separated by centrifugation after each washing, and the supernatants, containing inactive proteins, are discarded. Lysozyme ions eluted by washing the clay twice with 100 ml portions of 5% aq. pyridine solutions, which has been adjusted to pH 5 with sulfuric acid. The elutes are dialyzed against running tap water until no odour of pyridine remains. The volume increases considerably during the dialysis, which lasts for 24 hr. The pH of the final solutions is approximately 6. Amorphous, but essentially pure lysozyme is obtained by lyophillization of the solution. The yield is 0.7-0.8 g crystalline lysozyme or its salts are obtained from the amorphous material by one of the following procedures.

To 10 ml 5% solutions of amorphous lysozyme is added 0.5 g sodium hydrogen carbonate (final pH 8-8.5),and the solution is allowed to stand at room temperature until lysozyme carbonate crystallizes.

To 10 ml 5% solution of amorphous lysozyme is added 0.5 g sodium chloride, and the pH is adjusted to 9.5 to 10 with sodium hydroxide.

4.4   Isolation of Glutamine from Red Beet

Introduction

Glutamic acid is usually obtained as a secondary product during the hydrolysis of protein. The primary product is glutamine the amide of glutamic acid.

Materials

Bentonite,	Eggs,	Pyridine
Sodium chloride,	Sodium hydrogen carbonate	
Phosphate buffer,	Potassium chloride	

Procedure

Soak about 2 kg of red beet for 24 hr in 5 lit of 0.2 M ammonium sulphate solution. Wash them thoroughly with water and keep them overnight in a refrigerator. Slice the frozen beet and then ground in a meat mincer. Warm the pulp to room temperature and extract the juice with the help of cotton cloth in a 2 lit beaker. Wash the cloth and pulp with water to the total affluent add basic lead acetate solution; boil it while stirring for 30 min and filter. Wash the precipitate with 200 ml water. To the yellow clear filtrate add freshly prepared mercuric nitrate reagent and reflux the mixture for 3 to 4 hrs. Cool and add 1N NaOH solution until a permanent turbidity appears. Dilute it with 200 ml water and filter. To the suspension again add 10% NaOH solution so that the pH of the suspension is between 5 to 6 and then leave the suspension overnight. Collect the precipitate on Buchner funnel and wash it with water. Transfer the precipitate into a 500 ml conical flask, add minimum amount of water to it so that the precipitate is now in the form of suspension. Add 1 ml of 4N H_2SO_4 to the suspension and pass H_2S gas for 2 hrs with constant agitation.

Filter the precipitate on a Buchner funnel and wash with water. Heat the clear pale yellow filtrate for 20 min on water bath in order to remove last traces of H_2S gas. Neutralize the solution to approximately pH 6 by adding ammonium hydroxide solution. Add a pinch of decolorizing carbon, heat on water bath and filter. Concentrate the filtrate until a syrupy mass is obtained. Add about 10-15 ml of warm alcohol and cool the flask in a refregitor and keep it overnight. Filter the crystals of glutamine washed with cold alcohol and dry, M.P 254-256 °C.

4.5   Color Reactions of Proteins

- Biuret reagent
- Ninhydrin reagent
- Xantho protein reagent
- Millon's reagent

To 1 ml of protein solution add few ml of a 10 % NaOH solution and 1 drop of a 0.1% $CuSO_4$ solution. Mix well, and in case a pink to purple color has not yet developed, add another drop of $CuSO_4$ solution and continue until you get some color. Avoid adding excess copper, since it obscures the true color. It produces a red or violet color.

To 2 ml of protein solution add 1 ml of 0.1% Ninhydrin solution. Mix and boil for 1 min, and then cool.

To 1 mg of protein, add 1 ml of conc. HNO_3 and boil until the amino acids are dissolved. Now cool and add a few drops of 40% NaOH solution until the solution is slightly alkaline produce orange. This reaction is due to the nitration of the benzene ring in phenylalanine, thyrosine and tryptophene.

To 2 ml of protein solution, add few drops of Millon's reagent and boil in a water bath. If no color appears at first, continue to add drops of the Millon's reagent to produce red precipitation.

5

Lipids

5.1 Introduction

Lipids (fats) belong to a big group of natural organic compounds, insoluble in water but soluble in organic solvents such as: diethyl ether, petroleum benzene, chloroform, benzene, acetone etc. Lipids (fats) also belong derivatives of natural lipids and related with them compounds which retain properties of lipids. Lipids (fats) occur in all live organisms. In plants they are present first of all in seeds and flesh of fruit and in animal organisms in different organs or as separated adipose tissue.

Functions of lipids (fats)

Lipids (fats) are the most concentrated source of energy: from 1 g of fats 9 kcal are released.

Lipids (fats) are convenient and main source of spare material (they allow to take breaks between meals, during work they enable organism to function beyond thermal neutrality zone – body temperature keeping).

Accumulated in tissue fats protect against excessive heat emission, enables to adopt at low temperature, inside organism keeps organs in constant location and prevents their dislocation.

Accumulated in organism lipids are a warehouse of water, 30-50% of adipose tissue makes up water, combustion of 100 g of adipose tissue releases 107 g of water.

Mixed lipids (fats) of nourishment are a source of vitamins soluble in fats: A, D, E, K, and Essential Fatty Acids (vitamin F).

Fats in food save the balance of proteins of vitamins from group B.

They have high satiating value – they inhibit gastric juice secretion, raise flavor of dishes.

They fulfill function of building material, are ingredient of cell membranes and constitute element of composition of many hormones, cholesterol and important intracellular substances.

Nutritional requirements

Fats should provide 25-30% of energetic value of a daily nutritious ration for an adult. They ought to be fats unsaturated, not chemically hydrogenated, devoid of trans isomers.

Classification of lipids (fats)

In terms of chemical build lipids can be divided into:

Simple Lipids: Esters of fatty acids and alcohols

- Neutral fats – Triglycerides
- Waxes

Compound Lipids: Compounds contains apart from fatty acids and alcohols, other constituents as well.

- Phospholipids
- Glycolipids
- Other compound lipids

Derived Lipids: Derivatives of simple lipids and compound lipids. Obtained from hydrolysis of fats, usually contains an even number of carbon atoms and are straight chain derivatives.

- Fatty acids
- Fatty alcohols

- Hydrocarbons
- Fatty aldehydes
- Vitamins A, D, E, K

Saturated fatty acids (more important)

- Butyric acid C_3H_7COOH
- Caproic acid $C_5H_{11}COOH$
- Caprylic acid $C_7H_{15}COOH$
- Capric acid $C_9H_{19}COOH$
- Lauric acid $C_{11}H_{23}COOH$
- Myristic acid $C_{13}H_{27}COOH$
- Palmitic acid $C_{15}H_{31}COOH$
- Stearic acid $C_{17}H_{35}COOH$
- Arachidic acid $C_{19}H_{39}COOH$

Monounsaturated fatty acids (more important)

- Myristoleic acid   (9-tetra decenoic acid)
- Palmitoleic acid   (9-hexa decenoic acid)
- Oleic acid   (9-octa decenoic acid)
- Erucic acid   (13-docosenoic acid)

Polyunsaturated fatty acids - (Omega-3 and Omega-6)

- Linoleic acid (LA)   **(Omega-6)**
- Gamma-linolenic acid (GLA)   **(Omega-6)**
- Arachidonic acid (AA)   **(Omega-6)**
- Alpha-linolenic acid (LNA or ALA)   **(Omega-3)**
- Eicosapentaenoic acid(EPA)   **(Omega-3)**
- Docosahexaenoic acid(DHA)   **(Omega-3)**

Cyclic fatty acids

- Sterulic acid $\qquad$ $C_{18}H_{33}COOH$

- Chaulmoogric acid $\qquad$ $C_{17}H_{31}COOH$

- Garlic acid $\qquad$ $C_{17}H_{29}COOH$

Trans fats

Natural liquid vegetable oils are comprised mainly of unsaturated fatty acids. Trans fats are formed when manufacturers turn liquid oils into solid fats. Trans isomer has the same chemical formula as normal isomers (Cis isomers) except for the fact that its hydrogen atom is in a different spatial arrangement. During the hydrogenation of vegetable oil – a commercial process to harden oil for production of fats like shortening and hard margarine. As a result of this process, oils become semisolid, more stable at room temperature and more saturated. Trans isomers melt at 44 °C (111 °F). Cis (normal) isomers melt at 13 °C (55 °F). Hydrogenation increases the shelf life and flavor stability of foods.

Trans fats can be found in a list of foods including vegetable shortening, margarine, crackers, cereals, candies, donuts (doughnuts), baked goods, cookies, granola bars, French fries (chips), snack foods, salad dressings, fats, fried foods, and many other processed foods. Trans fats are found naturally in small quantities in some foods including beef, pork, lamb, butter and milk. Typical French fries have about 40% trans fats and many popular cookies and crackers range from 30-55% trans fats, donuts have about 35-40% trans fats. Trans fats are known to increase blood levels of low density lipoproteins (LDL), or "bad" cholesterol, while lowering levels of high density lipoproteins (HDL), known as "good" cholesterol. It can also cause major clogging of arteries, type 2 diabetes and other serious health problems and was found to increase the risk of heart disease.

Essential Fatty Acids (EFA), Omega-3 and Omega-6

Fatty acids are simply components of fats. There are two essential fatty acids. Essential means we need to get them from the diet because the body can't produce them. The first is alpha-linolenic (LNA or ALA, omega-3), and it belongs to the omega-3 parent. We can find alpha-linolenic in flax seed and flax oils, rape seed and rape oils, walnuts, hemp oil, cold pressed canola oil, wheat germ, dark green leafy vegetables. Linoleic acid (LA, omega-6) is the other essential, belonging to the omega-6 parent. We can found it in soy oil, sunflower seeds, safflower seeds, pumpkin seeds, sesame seeds, corn oil, and in most nuts. They exist also in other (non-essential) omega-3 and omega-6 fatty acids, which our body can produce from the two essential ones.

Non-essential omega-3 fatty acids include: docosahexaenoic (DHA, omega-3) and eicosapentaenoic (EPA, omega-3), which the body makes from alpha-linolenic (LNA or ALA, omega-3). They are contained first of all in food of sea origin (in fish i.e., mackerel, salmon, tuna, halibut, cod, herring, sardine, pilchard), linseed and rapeseed oils. Infants and children need DHA for proper brain growth from their diet (and breast milk can have DHA), so in that sense we should classify DHA as an essential fat for children. Non-essential omega-6 fatty acids include: gamma-linolenic and arachidonic which our body makes from linoleic acid (LA, omega-6). Fish oils contain the non-essential omega-3 fatty acids EPA and DHA. Since the body can make them from LNA, it follows that it isn't absolutely necessary to eat fish or take fish oil supplements but in all cases it is necessary to get LNA.

However, in certain cases the conversion from LNA to EPA or DHA is not adequate. That can happen if we don't get enough of the raw material LNA, or if we don't get enough of vitamins: niacin, B_6, C or enough zinc and magnesium, which all are needed in the conversion from LNA to DHA and EPA. Also, if the diet contains too much omega-6 fats in comparison to LNA, then the conversion is slowed down, so increasing dietary intake of fish in this case is highly desirable. The highest value and biological activity possess omega-3. The appropriate ratio of omega-6 to omega-3 should be (<5:1). The ratio of omega-6 to omega-3 (<5:1) has been based on the simple logic that this approximates that ratio in our cell membranes and also from evidence regarding food habits of Paleolithic people. Light, air and heat destroy EFA, so should be kept away from light, heat and air.

Role of EFAs

- They make up one of essential building materials of cells, maintain stability in cell division by protecting chromosomes.

- They take part in metabolism of cholesterol and its transport (over a half of cholesterol esters occurs in form of joint with linoleic acid what makes easier their distribution in organism, they decrease cholesterol level in blood).

- They inhibit aggregation of thrombocytes, cause expansion of blood cells including coronary ones, act anti-arrhythmically.

- They are precursors to biosynthesis of prostaglandins and prostacyclines.

- Omega-3 seems to help regulate the body's blood sugar levels, which helps keep hunger at bay. In the long term, it is believed that a diet rich in omega-3 might lower the risk of diabetes and obesity.

- A new study of teenagers has found that consumption relates to lower hostility rates in teenagers. Hostility has been shown to predict the development and manifestation of heart disease.

- They participate in transportation of water and electrolytes through biological membranes.

- They regulate expelling of sodium ions from organism.

- Facilitate the conversion of lactic acid to water to carbon dioxide thereby speeding up muscular fatigue recovery.

- Assist in hemoglobin production.

- Mediating the release of inflammatory substances from cells that may trigger allergic conditions.

- Inhibit tumor growth.

Deficiency of EFAs causes:

Inhibition of growth and decrease of mass increment.

- Dermal changes and hair losing.

- Liver and kidney degeneration.

- Male sterility and female miscarriages.

- Increased sensitiveness on allergic changes and bacterial infections.

- Drop of cardiac muscle tension (lower systole power, worse blood circulation, swellings).

5.2 Preparation of Azelaic Acid from Castor Oil

$$CH_2 - CH_2 - COOH$$
$$|$$
$$(CH_2)_3$$
$$|$$
$$CH_2 - CH_2 - COOH$$

Introduction

Azelaic acid can be prepared by oxidation of castor oil with nitric acid by oxidation of ricinoleic acid with nitric acid and alkaline permanganate by oxidation of methyl oleate with alkaline permanganate, by ozonisation of oleic acid and decomposition of the ozonide, by the Grignard reaction with 1, 7-heptamethylenemagnesium dibromide, by the oxidation of dihydroxy stearic acid with dichromate in sulfuric acid and by other methods.

Castor oil is the glycerylester of ricinoleic acid. It is abundant in nature and has many industrial applications. In the following experiment, castor oil is hydrolyzed with alkali, yielding crude ricinoleic acid, which is oxidized with potassium permanganate to give azelaic acid.

Materials

Castor oil,   Magnesium sulfate,   Potassium permanganate,

Ethanol,   Potassium hydroxide,   Sulfuric acid

Procedure

Fifty gm castor oil is added to a solution of 50 gm potassium hydroxide in 100 ml 95% ethanol. The mixture is placed in a 500 ml flask equipped with a reflux condenser and is boiled for 3 hr. The solution is then poured into 300 ml water and acidified by addition of a solution of 10 ml concentrated sulfuric acid in 30 ml water. The acid that separates is washed twice with warm water. The yield of crude oily ricinoleic acid thus obtained is 48 g.

The 48 g of ricinoleic acid is dissolved in 320 ml water containing 13 g potassium hydroxide. In a 2 lit round bottomed flask equipped with a power full mechanical stirrer are placed 135 g potassium permanganate and 1.5 lit water at 35 °C. The mixture is stirred to facilitate solution of the permanganate, and, if necessary, heat is applied to maintain the temp. at

35 °C. When the permanganate has completely dissolved, the alkaline solution of ricinoleic acid is added in a single portion with vigorous stirring. The temparature rises to about 75 °C, stiiring is continued for half an hr, or until a test portion added to water shows no permanganate color. To the mixture is now added a solution of 80 g concentrated sulfuric acid in 250 ml water.

The acid must be added slowly and carefully to prevent too rapid evaluation of carbon dioxide with consequent foaming. The mixture is heated on a steam bath for 15 min to coagulate the magnesium dioxide is placed in a 400 ml beaker and boiled with 200 ml water in order to dissolve any azelaic acid that may adhere to it. This mixture is filtered while hot, and the filtrate is added to the main portion. The combined filtrates are evaporated to a volume of about 800 ml and this solution is cooled in ice. The crystals that separate are filtered with suction, washed once with cold water, and dried. The yield is 14-16 g material of MP 95-106. The crude substance is dissolved in 240 ml boiling water, filtered with suction, and allowed to cool. The crystals are filtered, washed with water, and dried yield 11g, MP 104-106 °C.

Molecular formula: $C_9H_{16}O_4$

Molecular weight: 188

5.3   Isolation of Cholesterol from Gall Stones

Materials

Gallstones,   diethyl ether,   methanol

Procedure

Weigh 8 g of pulverized gallstones and place them in a 125 ml flask. Add 40 ml of diethyl ether and heat the mixture on a steam bath until the cholesterol is dissolved. Filter the brownish yellow solution through a fluted funnel, while it is still hot and add a little ether to replace that lost through evoparation.The brown residue that collects on the filter paper is Bilurubin, a bile pigment derived from hemoglobin. Dilute this filtrate with 40 ml methanol, add a little decolorizing agent and heat the mixture on a steam bath.

Preheat a funnel, and then filter the hot solution through a fluted filter paper. Reheat the greenish yellow filtrate and add just enough water, dropwise to make the solution cloudy. The solution is now saturated at the boiling point and cholesterol will crystallize upon cooling. Collect the crystals by vacuum filtration using a Buchner funnel. Wash the crystals with cold methanol and allow them to stand for a time in the open Buchner funnel to allow the solvent to evaporate. Weigh the dried cholesterol crystals.

5.4   Color Reactions of Cholesterol

Some of the major test solutions are:

- Libermann-Burchard Reagent

- Salkowski Reagent

- Iodine – Sulfuric acid Reagent

- Rosenheim Trichloro acetic acid reagent

I ml of a 0.1% solution of cholesterol in chloroform, add 5 drops of acetic anhydride and few drops of concentrated sulfuric acid.

Prepare a mixture of 10 ml concentrated sulfuric acid and 2 ml water. Mix well and allow cooling. Then evaporate to dryness in test tube 1 ml of 0.15 cholesterol solution in chloroform. After the chloroform is removed completely and the tube has cooled, add 5 ml of the prepared sulfuric acid mixture and few drops of 0.1 N iodine in KI.

To 1 ml of the 0.1% cholesterol in chloroform solution, add 1 ml of trichloro acetic acid solution (9:1; TCA: water).

6

Flavonoids

6.1 Introduction

Flavonoids constitute one of the most characteristic classes of compounds in higher plants. Many flavonoids are easily recognised as flower pigments in most angiosperm families (flowering plants). However, their occurence is not restricted to flowers but include all parts of the plant.

Flavonoids are widely distributed phenolic compounds in the plant kingdom. They occur in all parts of plants as complex mixtures of different components. Their structure, consisting of two hydroxy substituted aromatic rings joined by a three carbon link (a C_6-C_3-C_6 configuration) renders them hydrogen and electron donors. Thus, they are effective scavengers of free radicals, which are intermediate products of lipid peroxidation, and they slow down oxidation reactions. The flavonoids are polyphenolic compounds possessing 15 carbon atoms; two benzene rings joined by a linear three carbon chain.

The chemical structure of flavonoids are based on a C_{15} skeleton with a CHROMANE ring bearing a second aromatic ring B in position 2, 3 or 4.

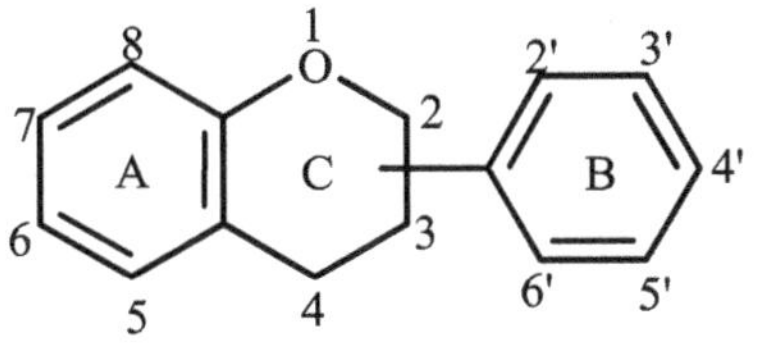

Various subgroups of flavonoids are classified according to the substitution patterns of ring C. Both the oxidation state of the heterocyclic ring and the position of ring B are important in the classification. Examples of the 6 major subgroups are:

Chalcones

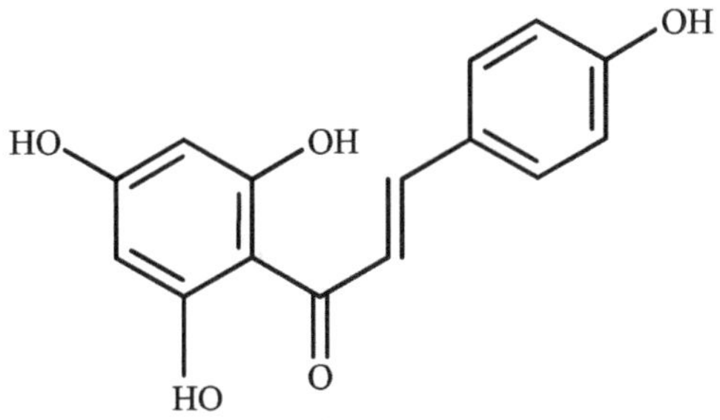

Flavone

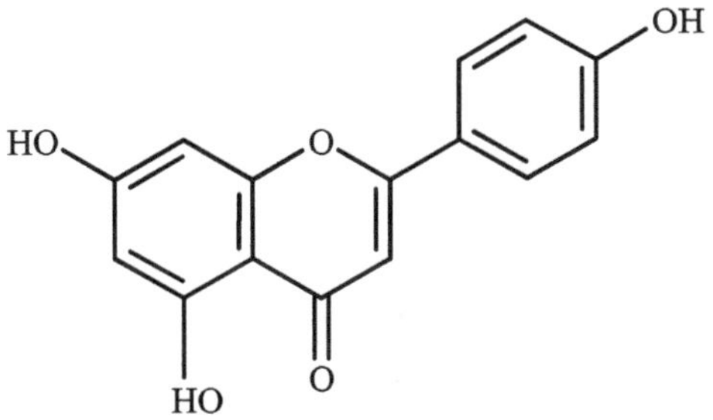

Flavonol

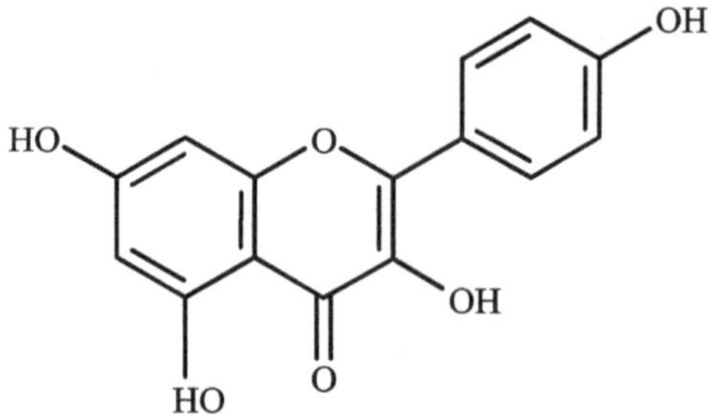

Flavanone

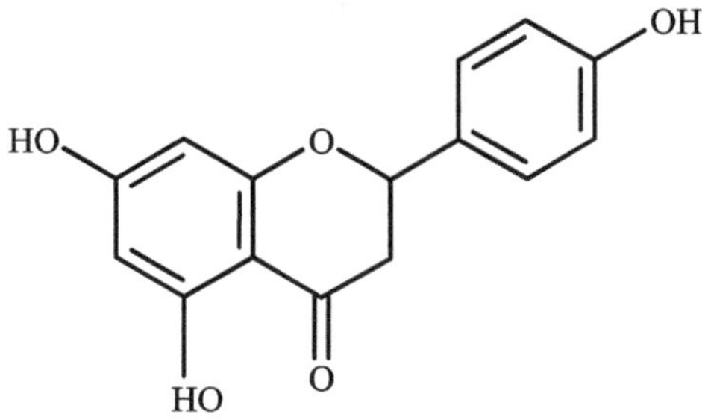

Anthocyanins

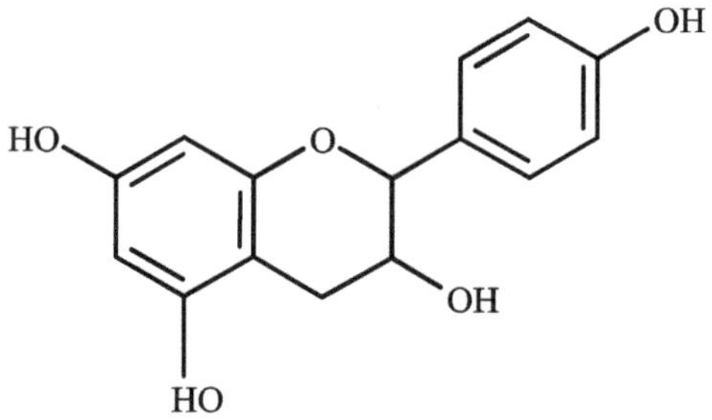

Isoflavonoids

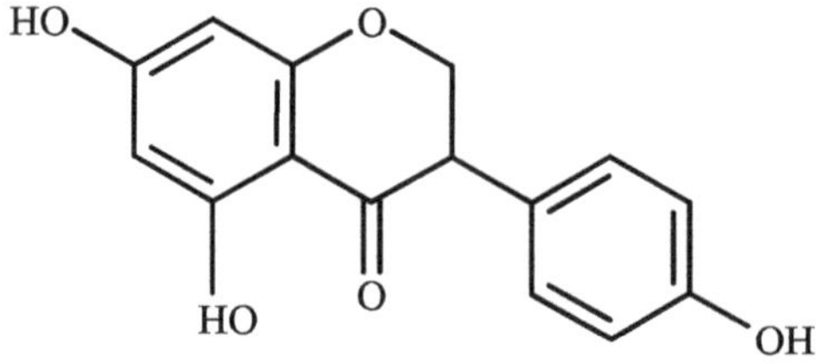

Most of these (flavanones, flavones, flavonols, and anthocyanins) bear ring B in position 2 of the heterocyclic ring.

In contrast to most other flavonoids, isoflavonoids have a rather limited taxonomic distribution, mainly within the *Leguminosae*. Most of our knowledge about the biosynthesis of isoflavonoids originates from studies with radioactive isotopes, by feeding labelled ^{14}C cinnamates. In isoflavonoids, ring B occupies position 3. The isoflavonoids are all colourless. It has been established that acetate gives rise to ring A and that phenylalamine, cinnamate and cinnamate derivatives are incorporated into ring B and C-2, -3, and -4 of the heterocyclic ring.

The isoflavonoids are all colourless. It has been established that acetate gives rise to ring A and that phenylalamine, cinnamate and cinnamate derivatives are incorporated into ring B and C-2, -3, and -4 of the heterocyclic ring.

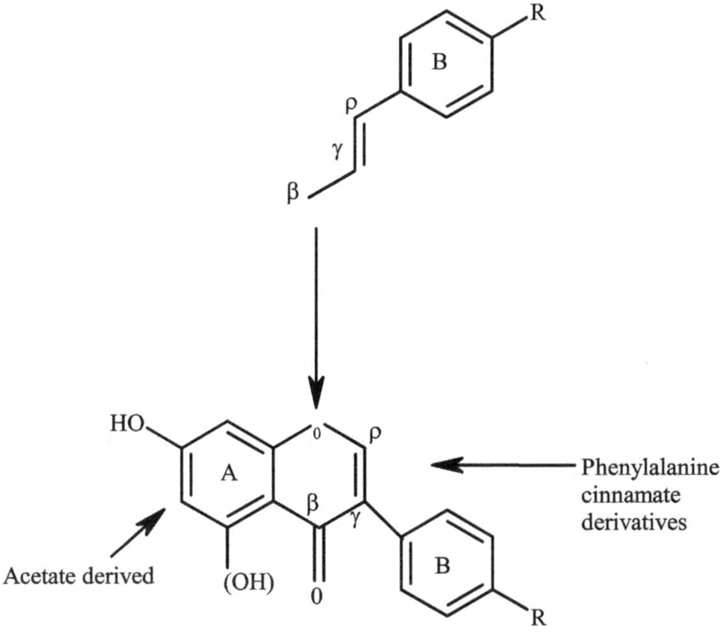

A group of chromane derivatives with ring B in position 4 (4-phenyl-coumarins = NEOFLAVONOIDS) is shown below.

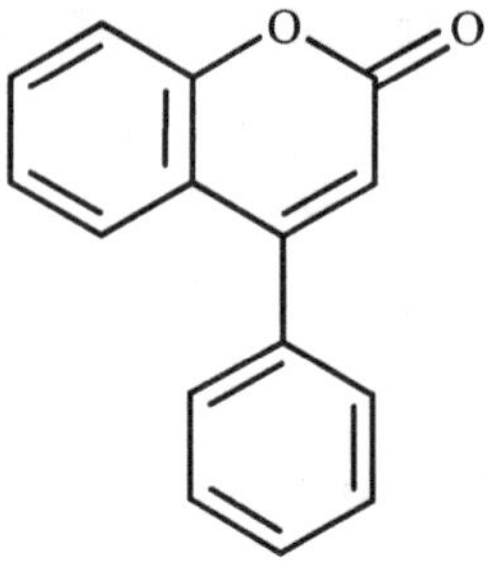

The Isoflavonoids and the Neoflavonoids can be regarded as *Abnormal Flavonoids*.

6.2   Isolation of Hesperidine from Orange Peel

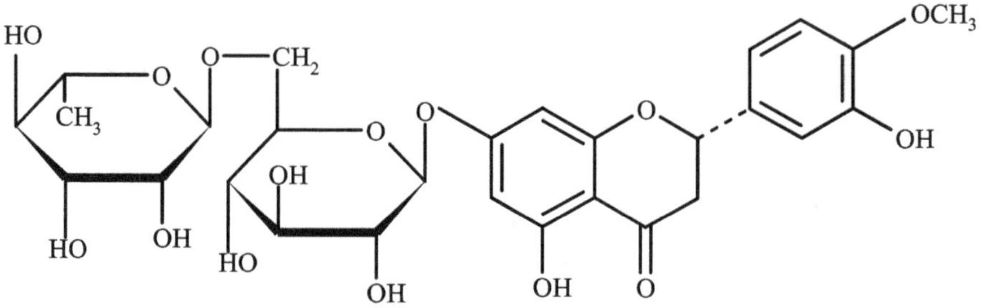

Introduction

Hesperidins was first isolated by Leberton in 1828 from the albedo (the spongy inner portion of the peel) of oranges of the family hesperides, and was given the name Hesperidine. The flavonoid hesperidin is a flavanone glycoside (glucoside) comprised of the flavanone (a class of flavonoids) hesperitin and the disaccharide rutinose. Hesperidin is the predominant flavonoid in lemons and oranges. The peel and membranous parts of these fruits have the highest hesperidin concentrations. Therefore, orange juice containing pulp is richer in the flavonoid than that without pulp. Sweet oranges *(Citrus sinensis)* and tangelos are the richest dietary sources of hesperidin. Hesperidin is classified as a citrus bioflavonoid. Hesperidine, in combination with a flavone glycoside called diosmin, is used in Europe for the treatment of venous insufficiency and hemorrhoids. Hesperidin, rutin and other flavonoids thought to reduce capillary permeability and to have anti-inflammatory action were collectively known as vitamin P. These substances, however, are not vitamins and are no longer referred to, except in older literature, as vitamin P.

Principle

Hesperidine can be isolated by two methods:

1. By extracting the dry peel successively with petroleum ether and methanol

2. By alkaline extraction of chopped orange peel and acidification of the extract.

Because of its highly insoluble, crystalline nature, hesperidine is one of the easiest flavonoids to isolate.

Materials

Acetic acid   Calcium hydroxide

Ferric chloride Formamide

HCl   Methanol

Orange peel   Ether isopropanol

Magnesium   Sodium borohydrate.

Procedure

Method A: Weight accurately 200 gm sun dried peel powdered in a disintegrator. The powder is placed in a 2 lit round bottomed flask attached to a reflux condenser. One litre of petroleum ether Bp 40-60 is added and heated on water bath for 1 hr. The contents of the flask are filtered while hot through a Buchner funnel, and the powder is allowed to dry at room temp. The dry powder is returned to the flask, and 1 lit of methanol is added. The contents are heated under reflux for 3 hr and then filtered hot and washed with 200 ml hot methanol. The filtrate is concentrated under reduced pressure leaving a syrupy residue crystallized from dilute acetic acid, yielding white needle; mp 252-254 °C.

Method B: The chopped orange peel 200 gm and 750 ml 10% calcium hydroxide solution are placed in a 2 lit round bottomed flask and thoroughly mixed, then left overnight at room temp. The mixture is filtered through a large Buchner funnel containing a thin layer of celite on the filter paper. The yellow orange filtrate is acidified carefully to pH 4-5 with concentrated HCl. Hesperidin separates as amorphous powder. It is collected on a Buchner funnel, washed with water, and recrystallized from aqueous formamide.

Acid degradation of Hesperidine:

On acid hydrolysis Hesperidine yields one mole of each of rhamnose, glucose and the aglycone hesperitin.

Chemical Test

Ferric chloride test: Addition of ferric chloride solution to hesperidine produces a wine red color.

Magnesium hydrochloride acid reduction: Dropwise addition of concentrated HCl to an ethanolic solution of hesperidin containing magnesium develops a bright violet color.

Spectral data:

^{13}CNMR (δ ppm):

C2	78.4	C1′	131.2
C3	42.0	C2′	114.3
C4	196.7	C3′	146.7

C5	163.0	C4′	148.1
C6	96.7	C5′	112.7
C7	165.2	C6′	117.8
C8	95.8	OMe	56.0
C9	162.5		
C10	103.5		

EI-Mass (m/z, %):

Molecular formule: $C_{28}H_{34}O_{15}$

Molecular weight: 610

593 ({M+H-H$_2$O}+, 3)

303 (100)

413(16)

465(34)

449(28)

6.3   Isolation of Naringin from Grape Fruit Peel

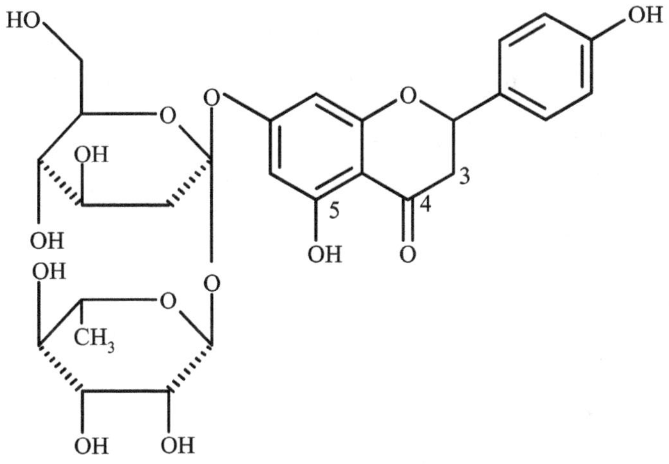

Introduction

The flavonoid of the albedo is called naringin and was discovered in 1866 by de Vry in the flowers of grapefruit trees of Java. It was also isolated from the Japanese bitter orange from leaves of pseudaegle trifoliate and from other sources. Naringin is extracted with hot water from the grapefruit peel along with a small quantity of pectin. Concentration of extract to about one ninth the original volume affords naringin, as an octahydride. Recrystallization from isopropanol gives the pure dehydrate.

Materials

Grapefruit peel,   Ethanol

Isopropanol,   Celite.

Procedure

Chopped grapefruit peel and four parts water are heated to 90 °C, and maintained at this temp. for five min. The water extract is filtered off. Two parts water are added to the solid; the extraction is repeated at 80 and immediately followed by filtration. The combined extracts are boiled with 1% celite, filtered, and concentrated in vacuo to approximately one ninth the original volumes. The concentrated extract is allowed to crystallize in a refrigerator and is then filtered as an octa-hydrate of mp83 (needles) naringin is dissolved in 100 ml boiling isopropanol and filtered hot. The filtrate is heated to its boiling point to initiate crystallization, then allowed to cool, filtered on a Buchner funnel, and washed with cold isopropanol; mp of dehydrate is 171 °C.

Spectral data:

^{13}CNMR (δ ppm):

C2	78.6	C1′	128.7
C3	42.0	C2′	128.0

C4	196.7	C3′	115.3
C5	162.9	C4′	157.1
C6	96.5	C5′	115.7
C7	164.9	C6′	128.2
C8	95.4		
C9	162.5		
C10	103.5		

EI-Mass (m/z, %):

Molecular formule: $C_{27}H_{32}O_{14}$

Molecular weight: 580

581(M + H, 100)

579 (M − H, 100)

7

Vitamins

7.1 Introduction

Vitamins are a group of organic compounds or micronutrients that are present in all living things. They are available in minute quantities of natural foodstuffs and are usually obtained from a normal diet. Vitamins are essential to normal metabolism, maintaining life and promoting growth and good health. Vitamins are considered micronutrients because the body needs them in relatively small amounts, in comparison with other nutrients, such as carbohydrates, proteins, fats and water. We have come to realize that vitamins not only sustain life, but also play an indispensable role in our daily well-being.

The first vitamin was discovered in 1897 by a Dutch biologist named Eijkman. He found that when bran was removed from rice, people consuming the refined rice developed beriberi, a serious disease. Eijkman also observed that when people ate the rice with the bran intact, no beriberi resulted.

This finding directed Eijkman and other scientists to chemically analyze rice for the substance which, when not present in adequate amounts, resulted in the development of beriberi. Thiamine, named vitamin B1, was discovered to be this mystery substance. Vitamin is the general term given to a group of organic substances that are present in food in minute quantities, but are distinct from carbohydrates, lipids, and proteins. They are essential for growth and cause specific deficiencies when not adequately supplied by the diet or improperly absorbed from food. So far, more than fifty vitamins have been identified.

Funk, in 1911, named this class of substances vitamine to indicate that these food factors were necessary for life, therefore, a vital amine. The final "e" was dropped when chemical analysis of these factors showed that not all of them were amines.

Vitamins are organic molecules that are necessary for normal metabolism in animals, but either are not synthesized in the body or are synthesized in inadequate quantities. Consequently, vitamins must be obtained from the diet. Most vitamins function as coenzymes or cofactors. There are vitamin-like substitutes that fail to meet all the criteria necessary to be classified as vitamins. They still have some properties of a vitamin and, in some cases, are present in larger amounts than vitamins. There are others that the body can synthesize in sufficient amounts to meet its needs if precursors are present. Some such substances include bioflavonoids, inositol, choline, PABA, and carnitine.

Vitamins must have the following five characteristics in order to be classified as such.

(i)   Required in relatively small quantities.

(ii) Essential because certain chemical reactions cannot occur without them.

(iii) Must be obtained in the diet because the body cannot manufacture them or cannot make adequate amounts.

(iv) Must be eaten regularly because they are stored in limited quantities and are gradually lost.

(v)   A deficiency results in at least one specific disorder.

Vitamins are of two distinct types

Water Soluble Vitamins

Vitamin C

Discovered originally in 1747 by Scottish naval surgeon, James Lind, who noticed that a "substance" in citrus fruits (ascorbic acid) prevented scurvy, it was re-discovered in 1912 by Hoist and Froelich. Scientist, Dr. Albert Szent-Gyorgyi, won the Nobel Prize for his discovery of vitamin C, and vitamin C was the first vitamin to be synthesized in 1935 in a process created Dr. Tadeusz Reichstein of the Swiss Institute of Technology in Zurich. Some claim that vitamin C is the most essential and beneficial vitamin.

It is a major antioxidant, strengthens blood vessels and fights infection, viruses and bacterial toxins. It is also a natural laxative and used in the formation of red blood cells. Vitamin C lowers the incidence of blood clots in veins and protects the brain and spinal cord from damage. It helps wounds heal and preserves and mends connective tissue. Ascorbic acid is essential for collagen production in the body and helps decrease cholesterol.

Vitamin B complex

They are said to be one of the most essential groups of vitamins and vital in maintaining the health of the nervous system, skin, eyes, hair, liver, brain function, muscle tone and gastrointestinal tract. These vitamins *together* are responsible for helping enzymes release energy from food, promote proper metabolism, give cells plenty of oxygen, detoxify organs, stabilize your nervous system functions, keep skin and hair healthy, prevent defective vision, and have also been used in the treatment of debilitating conditions.

Vitamin B₁ (Thiamine)

Discovered in 1912 by Casimir Funk, vitamin B_1 is said to be good for circulation, carbohydrate metabolism, cognitive activity, brain function and nervous system health.

Vitamin B₂ (Riboflavin) or (Vitamin G)

Discovered in 1926 by Tishler and Williams (also inventors of synthetic vitamins), it is needed for the digestion and metabolism of protein, fats and carbohydrates and is also necessary for cell respiration. Vitamin B_2 is required for the formation of red blood cells and antibodies and benefits vision, healthy skin, nails and hair. It also aids growth and reproduction and helps in stressful situations.

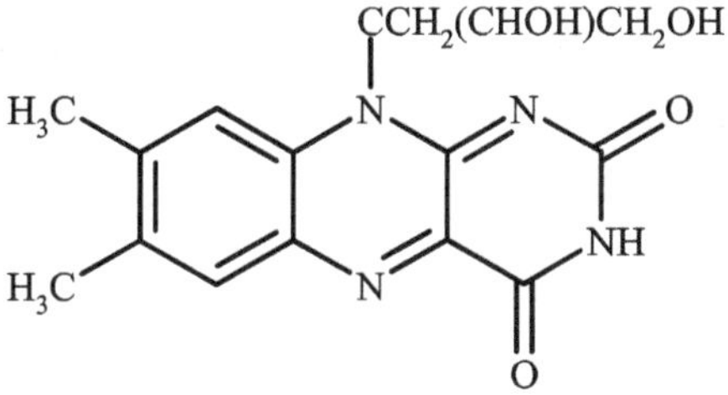

Vitamin B₃ (Niacin) or (Niacinamide or Vitamin P)

Discovered in 1937 by Conrad Elvehjem, niacin is said to be helpful in lowering serum cholesterol, reducing high blood pressure, preventing fatty buildup in the liver, maintaining the nervous system and helping to reduce depression. Vitamin B_3 is needed for proper circulation and healthy skin and thought to be better tolerated when taken with vitamin C (niacin has a tendency to cause flushing and a warm, tingling sensation).

Vitamin B₅ (Pantothenic Acid)

Vitamin B_5 is said to be the anti-stress vitamin and involved in the production of neurotransmitters. It also aids in vitamin utilization and helps to convert fats, carbohydrates and proteins into energy. Pantothenic Acid assists in cell building and required by all cells in the body. Vitamin B_5 improves the body's resistance to stress and assists in the development of the central nervous system. It is also said to help adrenal glands and to fight infections by building antibodies.

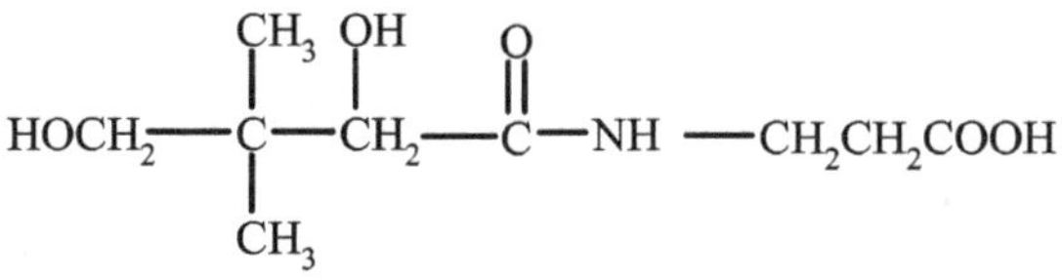

Vitamin B₆ (Pyridoxine) or (Pyridoxamine)

Discovered in 1934 by Paul Gyorgy, Pyridoxine is said to be good for menstrual problems (removing excess fluid) and for reducing the risk of arteriosclerosis and stroke. Vitamin B_6 aids in the formation of antibodies and in fat and carbohydrate metabolism. It is necessary for the body to produce lecithin (which is used clinically to clear out fatty livers and clogged arteries). Vitamin B_6 is necessary for the synthesis and breakdown of amino acids, the building blocks of protein. It also promotes healthy skin, reduces muscle spasms, carpal tunnel syndrome, leg cramps and numbing of the hands.

Vitamin B₇ (Biotin) or (Vitamin H)

After the initial discovery of biotin, nearly forty years of research were required to establish it as a vitamin. Biotin is required by all organisms but can be synthesized only by bacteria, yeasts, molds, algae and some plant species.

Vitamin B₉ (Folic Acid) or (Vitamin M)

Discovered in 1933 by Lucy Wills, folic acid is necessary for DNA and RNA synthesis, which, in turn, is essential for the growth and reproduction of all body cells. It is essential for the formation of red blood cells by its action on bone marrow and aids in amino acid metabolism.

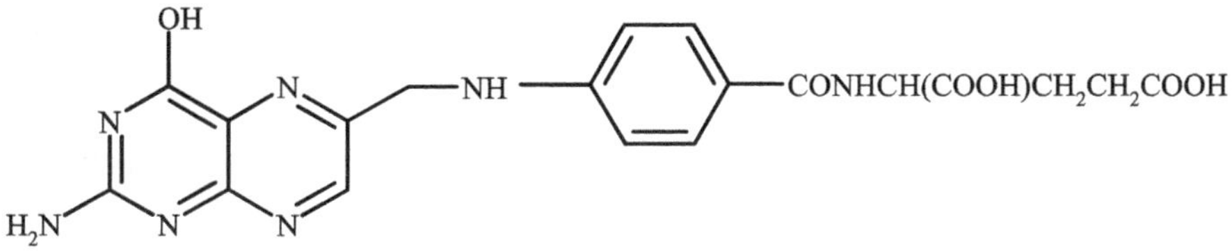

$$\text{H}_2\text{N} \ldots \text{(pteridine-PABA-glutamate structure)} \ldots \text{—CONHCH(COOH)CH}_2\text{CH}_2\text{COOH}$$

Vitamin B₁₂ (Cobalamin) or (Cyanocobalamin)

It is said to help in the formation and regeneration of red blood cells, helping to prevent anemia. It is necessary for calcium absorption and carbohydrate, fat and protein metabolism. Vitamin B_{12} helps to maintain a healthy nervous system, promotes growth in children and increases energy.

Fat Soluble Vitamins

Vitamin A

Discovered in 1912-1914 by McCollum and Davis, vitamin A was first synthesized in 1947. Benefits include enhancing the immune system and strengthening membranes, thus fighting infection. It is considered one of the antioxidant vitamins. Vitamin A is said to be good for eye problems and aids vision, as well as helping to keep our skin clear and reproductive system healthy. Other benefits include helping jawbone and tooth formation, maintaining healthy gums and combating otosclerosis (ear ailment). Overuse can be toxic and place unnecessary stress on the liver.

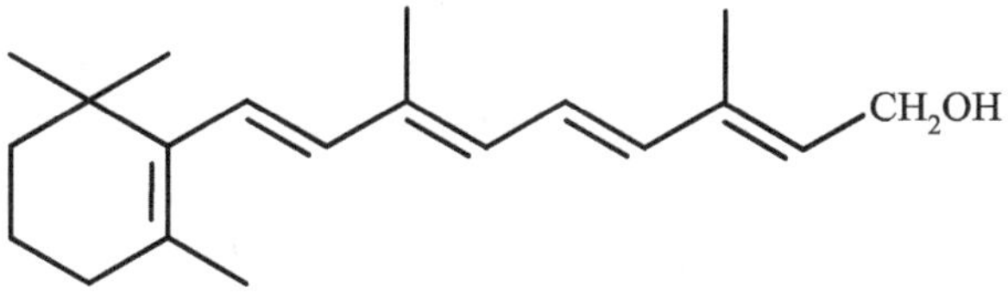

Vitamin D

Vitamin D was discovered in 1922 by Edward Mellanby and is sometimes called the "Sunshine" vitamin, because the sun is its source. Vitamin D helps the body utilize calcium and phosphorus, helping the body to form strong bones, teeth and healthy skin. It is necessary for thyroid function and growth, and it protects against muscle weakness. It is also involved in the regulation of the heartbeat and plays a role in preventing breast and colon cancer and osteoporosis. Excessive use can cause toxicity.

Vitamin E

Vitamin E was discovered in 1922 by Evans and Bishop. Vitamin E guarantees that the organs will be supplied with oxygen. It strengthens capillary walls, rejuvenates blood and helps skin lesions heal internally and externally. Vitamin E is considered a potent antioxidant and helpful in preventing cancer and cardiovascular disease. It is essential for reproduction and beneficial for male and female fertility and also prevents blood clots. Moreover, vitamin E supplies oxygen to the body for more endurance and is said to protect tissues from damage by environmental pollutants.

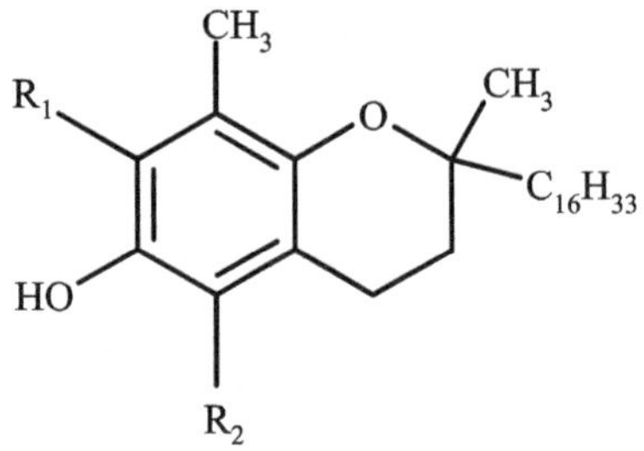

Vitamin K

Vitamin K is needed for the production of prothrombin, which is necessary for blood clotting and is essential for bone formation and repair. Vitamin K plays an important role in intestinal health and in converting glucose into glycogen for storage in the liver. It also may increase resistance to infection and certain cancers. This vitamin can be formed by natural bacteria in the intestines and is considered a vitality and longevity factor. Vitamin K is needed when taking antibiotics, because healthy, necessary bacteria are destroyed during this time by the activity of the antibiotics.

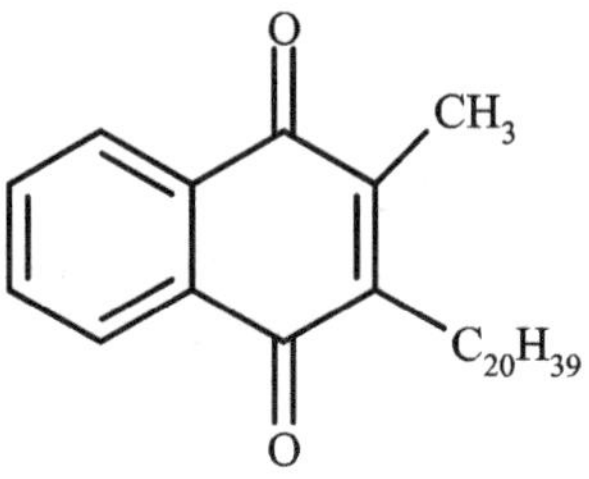

Vitamin	Major Source	RDA	Deficiency
Vitamin A	Animal tissues, especially fish and liver, green vegetables	700 µg for women 900 µg for men 770 µg for pregnant women 1,200 µg for breastfeeding women	Night blindness, Xerophthalmia, Bitot's spot
Vitamin D	Synthesized in the skin when exposed to sunlight. Also present at low concentration in some natural foods, and in many artificially-fortified food products.	200 IU for people aged 50 and younger 400 IU for people aged 51 to 70 600 IU for people older than 70	Rickets in children Osteomalacia in adults
Vitamin E	Vegetable oils, leafy green vegetables and whole grains.	15 mg (22 IU of natural or 33 IU of synthetic) 19 mg for breastfeeding women	Degenerative disorders of skeletal muscle, Muscle dystrophy
Vitamin K	Majority is synthesized by bacteria in the large intestine, photosynthetic (green) parts of plants	90 µg for women 120 µg for men	Heamorrhagic disorders
Vitamin C	Fruits and vegetables, citrus fruits, strawberries, tomatoes and leafy green vegetables.	75 mg for women 90 mg for men 85 mg for pregnant women 120 mg for breastfeeding women 35 mg more for smokers	Scurvey

Vitamin	Major Source	RDA	Deficiency
Vitamin B_1	Meats, leafy green vegetables, grains and legumes.	1.1 mg for women 1.2 mg for men 1.4 mg for pregnant or breastfeeding women	Beri-Beri
Vitamin B_2	Present in wide variety of foods, including milk, meats and grains.	1.1 mg for women 1.3 mg for men 1.4 mg for pregnant women 1.6 mg for breastfeeding women	Glossitis, Cheilosis
Vitamin B_3 Niacin	Meats, leafy green vegetables, potatoes and peanuts. Can be synthesized in small amounts within the body from tryptophan.	14 mg for women 16 mg for men	Pellagra, Dermatitis
Vitamin B_6	Present in meat from mammals, fish and poultry. Vegetables, potatoes and tomatoes.	1.3 mg for men 1.5 mg for women 1.9 mg for pregnant women 2.0 mg for breastfeeding women	Microcytic anemia, Homocystinuria, Cystathioninurea
Vitamin B_{12}	It is obtained almost exclusively from ingestion of animal products	2.4 µg for men 2.6 µg for pregnant women 2.8 µg for breastfeeding women	Macrocytic anemia

Vitamin	Major Source	RDA	Deficiency
Pantothonic acid	Grain, legumes, egg yolk, and meat. Also synthesized by intestinal bacteria.	5 mg (but no RDA has been established)	Alopecia, Degenerative changes in CNS
Biotin	Found in egg yolk, legumes, nuts and liver. Also synthesized by intestinal bacteria.	30 µg (but no RDA has been established)	Graying of hair, CNS disorders
Folic acid	Dark-green vegetables (spinach!), beef, eggs, whole grains. Also synthesized by intestinal bacteria	400 µg for women 600 µg for pregnant women 500 µg for breastfeeding women	Leucopenia, macrocytic anemia, neurological disorders

7.2   Isolation of Citric Acid from Lemon

$$\underset{\displaystyle \underset{\displaystyle CH_2 -\!\!-\!\!- COOH}{|}}{\overset{\displaystyle \overset{\displaystyle CH_2 -\!\!-\!\!- COOH}{|}}{HO -\!\!-\!\!- C -\!\!-\!\!- COOH}}$$

Introduction

In the 1600s, physicians became aware that daily intake of lemon juice would prevent outbreaks of scurvy among sailors on long sea voyages. Scurvy is a vitamin deficiency disease characterized by muscle wasting, inability of wound healing, bruising, and gum deterioration. Citrus fruits in general contain sugars, polysaccharides, organic acids, lipids, carotenoids (responsible for color), vitamins, minerals, flavonoids, limonoids (causing bitterness), and volatile components.

The lemon is a good source of potassium (145 mg/100 g of fruit), bioflavonoids, and vitamin C (40 to 50 mg/100 g, twice as much as oranges). The isolation of vitamin C from lemon juice has been performed. Calcium (61 mg) is also present, along with vitamins A, B_1, B_2, and B_3. The fruit is also low in calories, containing 27 Kcal/100 g.

The lemon is important for its nutritional value. Vitamin C is necessary to sustain the body's resistance to infection and heal wounds.

Materials

Lemons

NaOH

Calcium chloride.

Procedure

90 ml of thawed frozen lemon juice concentrate into 250 ml beaker and carefully add 10% NaOH solution with constant stirring until the mixture is slightly alkaline. A distinct color change occurs at this point, the solution

passing from a clear yellow to a brownish color. Strain the solution through muslin cloth to remove large particles of pulp and then filter through paper in a Buchner funnel. The pores of the filter paper may tend to become clogged by the extract in spite of the previous straining, change it as required to complete filtration. Place the filtrate in a beaker and add 5 ml of $CaCl_2$, stirring constantly for each 10 ml of filtrate.

Heat to boiling and filter off the copious precipitate of calcium citrate, from the hot solution using a Buchner funnel. Wash the precipitate with a small quantity of boiling water. Now resuspend it in a minimum quantity of cold water, heat to boiling, and once more collect the insoluble calcium citrate by filtration. Allow the salt to air dry. Citric acid may be prepared from the citrate salt by weighing the air dried salt and placing it in a beaker. Add sufficient 1N H_2SO_4 required to convert the salt to acid. Allow the mixture to stand for a few minutes, filter off the insoluble $CaSO_4$ and concentrate the filtrate to a small volume in a steam bath. Citric acid crystallizes out. Filter and dry.

Spectral data:

UV spectroscopy:

Solvent	λmax (nm)
0.001N HCl	207
Methanol	210.2

IR spectroscopy (cm-1):

Carboxyl OH	(str)	3600-2900
Hydroxyl OH (str)		3450
C – H (str)		2900 – 3000
C = O		1730
C – O – H		1400
C – O (str)		1220

^{1}HNMR spectroscopy (ppm):

COOH	4.97 (s)
$-CH_2 - C - (OH) (COOH) CH_2$	2.84 (dd)
s = singlet dd = doublet	

^{13}CNMR spectroscopy (ppm):

C-1 & C-5	173.47
C-2 & C-4	43.83
C-3	74.14
C-6	176.77

EI-Mass spectroscopy (m/z, %):

Molecular formula: $C_6H_8O_7$

Molecular weight: 192

192 (M+, 100)

8

Analysis of Fixed Oil or Fats

8.1 Determination of Saponification Value

Principle and Discussion

Saponification value is a number which is expressed in mg that indicates the amount of KOH, required to neutralize the free fatty acid and to saponify the acid present in 1 g of sample and substance. The fat or fixed oil consists of free fatty acids and triglycerides of fatty acids which allow to under go soap formation under endothermic conditions. In excess amount of alkali and the unreacted alkali is back titrated with equivalent strength.

The determination of saponification number is used as an aid in the detection of triglycerides of acids containing less than 16 or more than 18 carbon atom because the value of constant is inversely proportional to the molecular weight of acid present. It also helps in determination of adulteration either it is mineral oil or fixed oil. The determination of saponification value is based upon back titration of acid-base in which the excess alkaline is allowed to treat with the fatty acids of sample to saponify and the excess unreacted alkali is being titrated with equal strength acid. The back titration is performed to find out the amount of alkali that is neutralized by fatty acids and its esters.

Standardization of 0.5N KOH

3.15 g of oxalic acid (equal to 0.5N oxalic acid) was weighed accurately and dissolved in 100 ml of water. Pipette out 10 ml and titrated against 0.5N KOH towards end point using phenolphthalein as indicator. The end point was appearance of immediate permanent pale pink colour. The normality was calculated by using formula

$$N_1 \times V_1 = N_2 \times V_2.$$

Standardization of 0.5N HCl

Pipette out 20 ml of KOH and titrated with 0.1 N HCL until pink colour was disappeared. Repeat the titration until the value was concurrent. The indicator used was phenolphthalein. Normality was calculated by volumetric formula.

Procedure

Place 2 g of oil in a 250 ml round bottomed flask and add 25 ml of a 0.5 N alc. KOH solution. Insert into the neck of the flask by means of a perforated stopper an air condenser and heat the flask over a water bath for 30 min with constant shaking in order to achieve soap formation and hydrolysis of triglycerides. Then add 1 ml of phenolphthalein and titrate the excess of KOH with 0.5N HCL until the pink colour was disappeared. Now run a blank test, using exactly the same amount of 0.5 alc. KOH.

$$\text{Saponification value} = (\text{volume of KOH consumed} \times 0.02805 \times 1000)/\text{wt}$$

$$\text{Volume of KOH consumed by fat} = (\text{blank value} - \text{titer value}) \times \text{strength of HCl} / N.$$

N = normality of KOH

Saponification Value

Calculation

Standerdisation of 0.5 M HCl with Sodium Carbonate

$$Na_2CO_3 + 2HCl \longrightarrow 2NaCl + H_2O + CO_2$$

Molecular weight of sodium carbonate in grams = 106

106 gm of $Na_2CO_3 \rightarrow 2 \times 1000$ ml – 1M qHCl

Equivalent wt 1 mol of Na_2CO_3 is Equivalent to 2 mol of HCl

53	$\rightarrow$	1000	ml	$\rightarrow$	1M	HCl
5.3	$\rightarrow$	100	ml	$\rightarrow$	1M	HCl
2.65	$\rightarrow$	100	ml	$\rightarrow$	0.5 M	HCl
1.325	$\rightarrow$	50	ml	$\rightarrow$	0.5 M	HCl
0.6625	$\rightarrow$	25	ml	$\rightarrow$	0.5 M	HCl
0.3312	$\rightarrow$	12.5	ml	$\rightarrow$	0.5 M	HCl
0.0265	$\rightarrow$	1	ml	$\rightarrow$	0.5 M	HCl

$$\text{Molarity of HCl} = \frac{\text{Wt. of primary standard} \times \text{Exepected Molarity}}{\text{Burette Reading} \times \text{Equivalent factor}}$$

Each ml of 0.5 M HCl equivalent to 0.0265 gm of Na_2CO_3

$$\text{Saponication value} = \frac{(a - b) \times 28.05 \times \text{Exact Molarity}}{\text{Wt} \times \text{Expected Molarity}}$$

a = Burette Reading

b = blank

Standard Values

Almond oil	190-200
Cord liver oil	180-192
Castor oil	176-182
Cotton seed oil	193-195
Olive oil	190-195
Carnauba wax	80-95
Tolu balsam	154-220
Sesame oil	188-193
Peanut oil	188-195
Hydrginated vegetable oil	188-198
Corn oil	187-193

8.2   Determination of Iodine Value by Pyridine – Bromide

Principle

Iodine value is the number, which expressed in grams in quantity of halogens (I_2) calculated as iodine that absorbed by 100 grams of the substance under the described conditions. The iodine values are determined by iodometric titrations in which the KI is allowed to react with sample and liberated I_2 is back titrated with 0.01 sodium thiosulphate using starch mucilage as indicator. The blank value has to be performed without sample.

The iodine value can be determined by the following methods:

Method A (Iodine Monochloride Method)

In which the unsaturated (C = C) of fatty acid is allowed to react with iodine monochloride (I + Cl⁻) and the excess iodine monochloride is converted into iodine by addition of KI and titrated against sodium thiosulphate in usual range.

Method B (Pyridine Bromide Method)

In which the unsaturated fatty acid is allowed to react with pyridine bromide solution. The excess bromide is destroyed by addition of KI, the liberated I_2 is back titrated with $Na_2S_2O_3$.

Significance

It is the measure of the unsaturated compound present in the sample based upon the addition of halogen across. It reacts slowly. Hence the pyridine may be employed as a reagent in the official determination. Among the chemistry of fat analysis, it is the method to determine the degree of unsaturation in terms to find its purity with respect to rancidity, oxidative hydrolysis and hydrogenation etc. Most of the edible oil has iodine value range between 80-200. The low iodine value indicates the low degree of unsaturated and high iodine value indicates the high degree of unsaturation. High iodine value due to its more centers of unsaturation and as other hands solid fats exhibit low degree of I_2 value due to high preparation of esters of saturated fatty acid rather than the unsaturated fatty acid. E.g., Theobroma oil: 35-40

Procedure

Place an accurately weighed quantity of 0.2 gram of sample in a clean dry 500 ml iodine flask and add 10 ml of CCl_4 and dissolve. To this add 25 ml of pyridine bromide solution and immediately close with stopper of previously moistured with KI. Allowed this reaction mixture to stand in dark place at 25 °C for 30 minits, add 15 ml (2 grams) of HI solution in the top of iodine flask   and remove the stopper carefully and added 100 ml of water as side of the flask and shake well. Add 1 ml of starch mucilage and titrate against 0.01N $Na_2S_2O_3$ towards end point disappearance of blue colour as an end point.

The blank value was performed under the same condition and I_2 value was calculated from the following formula.

Iodine value: (Blank X N)/(weight taken $\times$ required molarity)

N = Normality of $Na_2S_2O_3$

Standard Values

Arachis oil   :   80-106

Rinsed oil   :   170-200

Castor oil   :   82-90

Cord-liver oil   :   150-180

8.3   Determination of Acid Value by Acid Base Titration

Principle and Discussion

Acid value is the number which is expressed in mg of KOH necessary to neutralize the free fatty acid present in 1 gm of sample or substance. The acid value is one of the chemical methods of analysis to perform on lipid in order to determine the amount of free acid present in the sample. It is applied in standardization of vegetable and medicinal oil in order to identify the standard and adulteration upon comparison with pharmacopoeial standard. Generally fixed oil consists of 70-90% of fatty acid in the form of ester as triglycerides and negligible amount of free fatty acid. Hence the higher acid value indicates the presence of more number of fatty acid and to declare the oil is impure or underwent hydrolytic oxidation that resulted the excess free fatty acids to give higher acid value.

Castor oil which is used as laxative and protective with acid value of not more 2 as per IP. The increased acid value indicates presence of impurity or long term storage. The acid value determination is based upon direct acid base titration in which the known quantity of sample or oil is titrated against 0.1M aqueous KOH in a suitable solvent like ethanol and ether using phenolphthalein as an indicator. Heating may be aided to enhance complete extraction of free fatty acids for that applied for solid fats.

Standardisation of 0.1N KOH

Pipette out 20 ml of 0.1N oxalic acid (0.63 gm in 100 ml) in a 250 ml clean dried conical flask and titrated against 0.1M KOH using phenolphthalein as an indicator towards end point. The end point was appearance of pale pink colour and the titration was repeated towards concordant values.

Molarity was calculated by using following formula

$$\text{Molarity of KOH} = V_1 \times N_1 / V_2$$

Procedure

Weigh accurately 2 gm of the substance to be examined in 10 ml mixture of 1:1 ratio of alcohol and solvent ether, which was previously neutralized by 0.1M KOH. If the sample does not dissolve in the cold solvent connect the flask with a reflux condenser and warm slowly with frequent shaking, until the sample dissolves. Add 1 ml of phenolphthalein solution and titrate with 0.1M KOH until the solution remains faintly pink after shaking for 30 sec.

$$\text{Acid value} = 5.61 \times n / wt$$

Acid Value

$$\text{Acid Value} = \frac{(a-b) \times 5.61 \times \text{Exact Molarity}}{\text{Wt} \times \text{Expected Molarity}}$$

Stadardisation of 0.1m KOH with Oxalic Acid

$$\begin{matrix} \text{COOH} \\ | \\ \text{COOH} \end{matrix} .2\,H_2O \; + \; 2KOH \longrightarrow \begin{matrix} \text{COOK} \\ | \\ \text{COOK} \end{matrix} \; + \; 2H_2O$$

1 mole of Oxalic acid is equivalent to 2 moles of KOH

Molecular weight of Oxalic acid in grams = 126

Equivalent wt 1 mol of Oxalic acid ($C_2H_6O_6$)

63	$\rightarrow$	1000	ml $\rightarrow$ 1M	KOH
6.3	$\rightarrow$	1000	ml $\rightarrow$ 0.1M	KOH
0.63	$\rightarrow$	100	ml $\rightarrow$ 0.1M	KOH
0.1575	$\rightarrow$	25	ml $\rightarrow$ 0.1M	KOH
0.0063	$\rightarrow$	1	ml $\rightarrow$ 0.1M	KOH

Each ml of 0.1 M KOH equivalent to 0.0063 gm of Oxalic acid

$$\text{Molaty of KOH} = \frac{\text{Wt of primary standard} \times \text{Exepected Molarity}}{\text{Burette Reading} \times \text{Equivalent factor}}$$

Standard Values

Castor oil	:	not more than 2
Oleicacid	:	196-204
Tolu balsam	:	112-168
White wax	:	17-24
Rosine	:	150-180

8.4   Determination of Peroxidee Value

Principle

The peroxide value is the number of milli equivalent of active Oxygen that express the amount of peroxide content in 1000 gm of substance. Peroxide value can be calculated from following formula according to I.P and U.S.P.

$$\text{Peroxide Value} = \frac{(a-b)\times 10}{Wt}$$

Peroxide value indicates amount of free fatty acid present in oil/fat. Upon long storage conditions simple lipids gets hydrolyzed to free fatty acids .These free fatty acids react with oxygen present in air and are converted to hydro peroxides. In acidic media hydroperoxides of free fatty acids acts as oxidising agents and oxidises potassium Iodide to Iodine and one themselves reduced to hydroxyl fatty acids .Iodine liberated in above conditions was titrated with 0.01M $Na_2S_2O_3$, i.e Peroxide value is determined by Iodometric titration.

Peroxide value is used to measure the degree of rancidity. Peroxide oxidise ferrous ion to ferric ions. Ferric ions react with potassium thiocyanate and form red colour ferricthiocyanate.

OIL(Sample) + Fe^{+2} $\longrightarrow$ **Fe^{+3} + Hydroxy fatty acids**

         Ferrous   **Ferric**

Fe^{+3} + Potassium thiocyanate $\longrightarrow$ **Fe(SCN)$_3$**

                                               (Red Colour)

Standardisation of 0.01m Sodium Thio Sulphate with Potassium Dichromate

It is an example of redox titration. Sodiumthiocyanate is a secondary standard. Its equivalent weight and molecular weight are same. It can be standerdised by using strong oxidizing agents like Pot. Iodide and Pot. dichromate.

Calculation

Molecular weight of Sodium thiosulphate $Na_2S_2O_3$ in grams =248.18

248.18	$\rightarrow$	1000	ml	$\rightarrow$	1M
24.818	$\rightarrow$	1000	ml	$\rightarrow$	0.1M
2.4818	$\rightarrow$	100	ml	$\rightarrow$	0.01 M
0.24818	$\rightarrow$	100	ml	$\rightarrow$	0.01M

$$K_2Cr_2O_7 = 3I_2 = 6 \times Na_2S_2O_3$$

294.18 gm of $K_2Cr_2O_7 - 6 \times 1000$ ml of 1M $Na_2S_2O_3$

49.038	$\rightarrow$	1000	ml	$\rightarrow$	1M
4.903	$\rightarrow$	1000	ml	$\rightarrow$	0.1 M
0.4903	$\rightarrow$	1000	ml	$\rightarrow$	0.01 M
0.04903	$\rightarrow$	100	ml	$\rightarrow$	0.01 M
0.024515	$\rightarrow$	50	ml	$\rightarrow$	0.01 M
0.01225	$\rightarrow$	25	ml	$\rightarrow$	0.01 M
0.00612895	$\rightarrow$	12.5	ml	$\rightarrow$	0.01 M
0.0004903	$\rightarrow$	1	ml	$\rightarrow$	0.01M

Each ml of 0.01M $Na_2S_2O_3$ = 0.0004903 gm of $K_2Cr_2O_7$

Procedure

5 gm of sample in a glass stoppered conical flack. Add 30 ml minor of 3 volumes of Glacial acetic acid and 2 volumes of Chloroform. Dissolve the oil by through stirring; add 0.5 ml saturated Potassium Iodide solution. Allow to standard for 1 min in dark, add 30 ml of distilled water by shaking.

Then add 0.5 ml of 10% starch solution, titrate with 0.01 M sodium thiosulphate until the end point blue to colourless.

$$\text{Peroxide Value} = \frac{(b-a) \times \text{E.factor} \times 1000 \times \text{Expect Molarity}}{\text{Wt} \times \text{Expected Molarity}}$$

Standard value

Arachis oil   NMT 5

Castor oil   NMT 5

8.5 Determination of Aldehyde Content in Cinnamon

Cinnamon oil is the volatile oil obtained from cinnamon zeylanicum. The main active constituent of cinnamon oil is cinnamaldehyde .Determination of aldehyde content in cinnamon oil depends up on the interaction of aldehydes with hydroxylamine HCl, and the formation of aldoxime and liberation of equivalent amount of HCl, which is determined with the standardized 0.5 M alc KOH using methyl orange as indicator.

Stadardisation of 0.1m KOH With Oxalic Acid

Molecular weight of KOH in grams = 56

56	$\rightarrow$ 1000 $\rightarrow$	ml	$\rightarrow$ 1M
5.6	$\rightarrow$ 100 $\rightarrow$	ml	$\rightarrow$ 1M
2.8	$\rightarrow$ 100 $\rightarrow$	ml	$\rightarrow$ 0.5M

1 mole of Oxalic acid is equivalent to 2 moles of KOH

Molecular weight of Oxalic acid in grams = 126

Equivalent wt 1 mol of Oxalic acid ($C_2H_6O_6$) is Equivalent to 2 mols of KOH

63	$\rightarrow$ 1000ml	$\rightarrow$ 1M	KOH
6.3	$\rightarrow$ 100ml	$\rightarrow$ 1M	KOH
3.15	$\rightarrow$ 100ml	$\rightarrow$ 0.5M	KOH
1.575	$\rightarrow$ 50 ml	$\rightarrow$ 0.5M	KOH
0.7875	$\rightarrow$ 25 ml	$\rightarrow$ 0.5M	KOH
0.0315	$\rightarrow$ 1ml	$\rightarrow$ 0.5M	KOH

Each ml of 0.5M alc KOH equivalent to 0.0315gm of Oxalic acid.

$$\text{Molarity of KOH} = \frac{\text{Wt of primary standard} \times \text{Expected Molarity}}{\text{Burette Reading} \times \text{Equivalent factor}}$$

Procedure

Weigh accurately 1g of cinnamon oil and transfer in to a cleaned glass stoppered flask. To this add 5ml of benzene or toluene and 15ml of 60% alc hydroxylamine HCl solution. Shake vigorously, the mixture and then add few drops of methyl orange indicator. Then titrate with 0.5M alc KOH until the red color changes to yellow.

Calculation

Equivalent factor determination:

1M Cinnamic aldehyde $\equiv$ 1M of Hydroxylamine HCl $\equiv$ 1M KOH

132 gm of Cinnamic aldehyde in 1000ml $\rightarrow$ 1M KOH

66 gm of Cinnamic aldehyde in 1000ml $\rightarrow$ 0.5M KOH

6.6 gm of Cinnamic aldehyde in 100ml $\rightarrow$ 0.5M KOH

0.066 gm of Cinnamic aldehyde in 1ml $\rightarrow$ 0.5M KOH

Each ml of 0.5M alc KOH $\equiv$ 0.066 gm of Cinnamic aldehyde

Aldehyde content:

$$\frac{\text{Vol of KOH consumed} \times \text{Exact Molarity} \times \text{factor}}{\text{Exepected Molarity}}$$

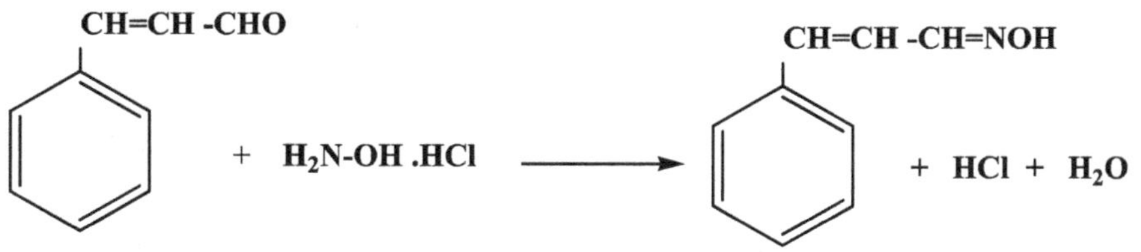

Rinse undissolved portion of the oil in graduated part of the neck of the flame by the gradual addition of more of the 5% KOH solution. Allow to stand for not less than 24 hours at room temperature and then read at the volume of the undissolved portion of the oil which measures between 1.0 and 1.5 ml.

8.6   Extraction and Isolation of Cinnamon Oil from Cinnamom Zeylancum

Introduction

The genus *Cinnamomum* comprises several hundred species, which occur in Asia and Australia. These are evergreen trees and shrubs and most of the species are aromatic. *C. zeylanicum*, the source of cinnamon bark and leaf oils, is a tree indigenous to Sri Lanka. Many species of cinnamon yield a volatile oil on distillation. The most important cinnamon oils in world trade are those from *C. zeylanicum*, *C. cassia* and *C. camphora*. The other species provide oils, which are utilized as sources for chemical isolates. However, a number of other cinnamon species are distilled on a much smaller scale and the oils used either locally or exported. Major compounds present in stem-bark oil and rootbark oil are cinnamaldehyde (75%) and camphor (56%), respectively (Senanayake *et al.*, 1978). Cinnamon bark oil possesses the delicate aroma of the spice and a sweet and pungent taste. It is employed mainly in the flavouring industry where it is used in meat and fast food seasonings, sauces, pickles, baked goods, confectionery, cola-type drinks, tobacco flavours and in dental and pharmaceutical preparations. Perfumery applications are far less than in flavours because the oil has some skin-sensitizing properties, so it has limited use in some perfumes.

The official method for the volatile oil in a crude drug is hydro distillation based on distilling the drug with water and the distillate is collected ion the graduated tube, from which the aq. portion of the distillate automatically returns to distillation tank.

Procedure

Weigh accurately 20gm of the crude powderd drug in 1litre distillation flask together with 250ml of water. Add few pieces of porcelain to it in order to avoid bumping during distillation. keep the distillation flask on sand bath and set the distillation assembly.fill the graduated receiver with water without any air bubbles; do not tighten the outlet near the lower end of the receiver instead loosly pack it with cotton. Turn the heat on continuos the distillation for 4hr at s rste which keeps the lower end of the condenser cool. Allow the distillate to be collected in the graduated receiver, in which the aq.portion of the distillate is automatically separated and return to the distillation flask. Measure the volumeof volatile oil which separate act as the upper layer in the graduated and calculate % v/w on dry weight basis.

8.7 Extraction of Clove Oil From Clove Buds

Introduction

It is a natural analgesic and antiseptic used primarily in dentistry for its main ingredient eugenol. It can also be purchased in pharmacies over the counter, as a home remedy for dental pain relief, mainly toothache; it is also often found in the aromatherapy section of health food stores. The oil produced by cloves can be used in many things from flavouring medicine to remedies for bronchitis, the common cold, a cough, fever, sor throat and tending to infections. The main oil-producing countries are Madagascar and Indonesia.

There are three types of clove oil:

Bud oil is derived form the flower-buds of *S.aromaticum*. It consists of 60-90% eugenol, eugenyl acetate, caryophyllene and other minor constituents.

Leaf oil is derived from the leaves of *S.aromaticum*. It consists of 82-88% eugenol with little or no eugenyl acetate, and minor constituents.

Stem oil is derived from the twigs of *S.aromaticum*. It consists of 90-95% eugenol, with other minor constituents.

Procedure

The plant (100 g of dried and ground clove buds) in 500 ml flask was submitted to hydrodistillation for 4–6 hr and steam distillation for 8–10hr. The volatile distillate was collected until no oil drop out. The distillate was saturated with sodium chloride and added with some ether. Then, the ether layer and hydro layer were separated by funnel. After dehydrated by anhydrous sodium sulphate, the ether layer was further heated in 60^0 C water bath to make oil to be concentrated and the ether to be recovered. The oil was weighed and refrigerated prior to analysis contentin clove oil.

9

Estimation of Nitrogen
(Kjeldahl's Methods)

Introduction

A known weight of nitrogen containing compound is digested in kjeldahl flask with concentrated sulphuric acid in the presence of a catalyst until the organic matter has been completely oxidized and the nitrogen converted quantitatively to ammonium sulphate.The acidic reaction mixture is then treated with excess of NaOH solution. The liberated ammonia is steam distilled and absorbed into saturated boric acid solution. Ammonium borate so formed is titrated with standard HCl solution using methyl red or methylene blue as indicator. From the amount of ammonia evolved, the percentage of nitrogen can be calculated.

Procedure

Weight accurately 30-50 mg of nitrogenous compound and then transfer into a clean and dried Kjeldahl flask. Add 1 gm of catalyst and 5-10 ml of concentrated sulphuric acid to it. Keep the flask loosely stoppard and then heat it in an inclined position slowly (until the frothing subsides) over the micro burner and then vigorously for 1-2 hr or until the reaction mixture is colorless. Transfer the reaction mixture to distillation flask, wash the kjeldahl flask 4-5 times with distilled water and transfer each washing to the distillation flask.

Fit the distillation flask with dropping funnel and kjeldahl trap. Connect the kjeldahl trap with condenser with the lower end dipping in 25 ml saturated boric acid solution contained in 100 ml conical flask. Introduce 30-40 ml of standard NAOH solution from the dropping funnel and heat the dilute reaction mixture inside the distillation flask by passing steam. Ammonia evolved is absorbed in to boric acid solution. The amount of ammonia evolved is obtained by titrating with standard HCl using methyl red or methylene blue as indicator.,

Calculation

$$\text{Percentage of nitrogen} = \frac{\text{Wt. of nitrogen}}{\text{Wt. of nitrogen sample}} \times 100$$

$$\text{Wt. of Nitrogen} = \frac{14 \times N \times V}{1000}$$

N = Normality of HCl

V = Volume of HCl used to neutralize ammonia

10

Estimation of Carbonyl Group

Introduction

The determination of percentage purity of carbonyl compound is applicable to both aldehydes and ketones by hydroxylamine HCl method. A known quantity of carbonyl compound is titrated with excess of hydroxylamine HCl in the presence of pyridine to yield oxime and pyridine HCl. The equilibrium of the reversible reaction is displaced to the right due to the presence of pyridine and hydroxylamine HCl. Pyridine HCL formed during the reaction is acidic enough to be titrated with NaOH solution using Bromophenol blue as an indicator.

Procedure

To 30 ml of hydroxyl amine HCl in a 250 ml conical flask add 1 gm of accurately weighed carbonyl compound followed by 100 ml of Bromophenol blue indicator. Stopper the flask and keep it at laboratory temperature for half an hour. Meanwhile carryout the blank titration using 30 ml of hydroxyl amine HCl and 100 ml of bromophenol blue indicator with methanolic sodium hydroxide solution to blue green color. Keep the colored solution aside for color comparison. Now titrate the reaction mixture for HCl liberated during the reaction with methanolic sodium hydroxide solution until the color matches with that of the blank.

Note : In case of Benzil or Acetophenone reflux on water bath.

Calculation

$$\% \text{ of Carbonyl compound} = \frac{a - b \times N \times M.\,Wt.}{W \times 1000} \times 100$$

N = normality of NaOH

M.Wt. = Molecular weight of sample

W = weight of the sample

a = Burett reading

b = blank

11

Estimation of Hydroxyl Group

Introduction

Estimation of hydroxyl group includes esterification and oxidation. Esterification process is applicable to nearly all compounds where as oxidation method is generally applicable to compounds containing vicinal hydroxyl groups. The most common esterification method is acetylation and it involves heating the known volume of alcohol with known volume of acetic anhydride in excess and pyridine until acetylation complete. During this process the OH group is replaced by acetyl group. Excess amount of unreacted acetic anhydride present in the reaction mixture is hydrolysed to acetic acid by adding water to it. The total free acetic acid is then estimated by titrating with standard NaOH solution using Phenolphthalein as an indicator. Blank titration is also performed under similar conditions.

Procedure

Place accurately weighed 1 gm of the alcoholic substances (menthol benzyl alcohol and cyclohexanol etc.) in a 100 ml round bottomed flask fitted with a reflux condenser. Add 20 ml acetylating mixture to the flask and heat it on water bath for about 1 hr. Remove the heat source and pour about 20 ml distilled water down the condenser and the shake the flask thoroughly to ensure complete hydrolysis of the residual acetic anhydride. Remove the condenser, cool the flask and keep it at room temperature for 5-10 min. Titrate the contents of the flask with standard 1N NaOH solution using phenolphthalein indicator. Perform the blank titration

Calculation

1000 ml 1N NaOH = 1 gm mol NaOH = 1gm mol acetic acid = 1 hydroxyl group

$$\text{Number of OH groups} = \frac{\text{Vol of NaOH}}{1000} \times \frac{\text{M.Wt}}{\text{Wt of sample}}$$

$$\text{Vol of NaOH} = (\text{Burett reading} - \text{Blank})$$

12

Interpretation of IR Spectroscopy

The infrared region of the electromagnetic spectrum extends from 14,000 cm^{-1} to 10 cm^{-1}. The region of most interest for chemical analysis is the mid-infrared region (4,000 cm^{-1} to 400 cm^{-1}) which corresponds to changes in vibrational energies within molecules. The far infrared region (400 cm^{-1} to 10 cm^{-1}) is useful for molecules containing heavy atoms such as inorganic compounds but requires rather specialized experimental techniques.

Molecular Vibrations

All molecules vibrate, even at a temperature of absolute zero. In general, a polyatomic molecule with N atoms has 3N–6 distinct vibrations. Each of these vibrations has an associated set of quantum states and in IR spectroscopy. The IR radiation induces a jump from the ground (lowest) to the first excited quantum state. Although approximate; each vibration in a molecule can be associated with motion in a particular group.

A simple example is methanal, whose 6 vibrations involve the following motions:

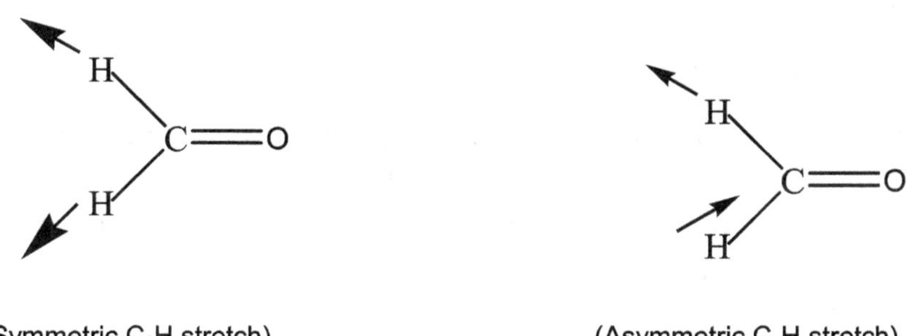

(Symmetric C-H stretch)　　　　(Asymmetric C-H stretch)

(C = O stretch)

(CH2 bend)

(HCO bend)

(Out-of-plane)

Clearly, even for this simple molecule, a description of the vibrations is quite complicated and this complexity rapidly increases as the size of the molecule increases.

The Fingerprint Region

The fact that there are many different vibrations even within relatively simple molecules, means that the infrared spectrum of a compound usually contains a large number or peaks, many of which will be impossible to confidently assign to vibration of a particular group. Particularly notable is the complex pattern of peaks below 1500 cm^{-1} which are very difficult to assign.

However, this complexity has an important advantage in that it can serve as a *fingerprint* for a given compound. Consequently, by referring to known spectra, the region can be used to identify a compound.

Interpretation of IR Spectra

To obtain a more detailed interpretation of an IR spectrum it is necessary to refer to correlation charts and tables of infrared data. There are many different tables available for reference and a brief summary is given below for some of the main groups. When assigning peaks to specific groups in the infrared region it is usually the stretching vibrations which are most useful. Broadly speaking, these can be divided into four regions.

Type	Region (cm^{-1})
Single bonds to hydrogen	3700 – 2500
Double bonds	2300 – 2000
Triple bonds	1900 – 1500
Single bonds (other than hydrogen)	1400 – 650

IR Absorption Frequencies of Some Organic Functional Groups:

Functional group	Absorption (cm^{-1})
Alcohol	
OH (str)	3200 – 3600
C-O (str)	1050 – 1150
Alkane	
C-H (str)	2850 - 3000
-C-H (bend)	1350 – 1480
Alkenes	
=C-H (str)	3010 – 3100
=C-H (bend)	675 – 1000
C=C (str)	1620 – 1680
Alkynes	
C – H(str)	3300
-C≡C-(str)	2100 – 2260
Alkyl halides	
C-F (str)	1000 – 1400
C-Cl (str)	600 – 800
C-Br (str)	500 – 600
C-I (str)	500
Amine	
N-H (str)	3300 – 3500
C-N (str)	1080 – 1360
N-H (bend)	1600
Aromatic	
C-H (str)	3000 – 3100
C=C (str)	1400 – 1600

Carbonyl

    C=O (str)   1670 – 1820

Acid

    C=O (str)   1700 – 1725

    O-H (str)   2500 – 3300

    C-O (str)   1210 – 1320

Aldehyde

    C=O (str)   1740 – 1720

    =C-H (str)   2820 – 2850 &2720 – 2750

Amide

    C=O (str)   1640 – 1690

    N-H (str)   3100 – 3500

    N-H (bend)   1550 – 1640

Anhydride

    C=O (str)   1800 – 1830

Ester

    C=O (str)   1735 – 1750

    C-O (str)   1000 – 1300

Ether

    C-O (str)   1000 – 1300

Nitrile

    CN (str)   2210 – 2260

Nitro

    N-O (str)   1515 – 1560 & 1345 – 1385

    α, β – unsaturated (str)   1665 – 1685

Cyclic rings

    3-membered   1850

    4-membered   1780

    5-membered   1745

    6-membered   1715

    7-membered   1705

    Arylketone   1680 - 1700

13

Interpretation of H^{1}NMR Spectroscopy

Chemical Shifts in H^{1}NMR Spectra

The signal frequency that is detected in nuclear magnetic resonance (NMR) spectroscopy is proportional to the magnetic field applied to the nucleus. This would be a precisely determined frequency if the only magnetic field acting on the nucleus was the externally applied field. But the response of the atomic electrons to that externally applied magnetic field is such that their motions produce a small magnetic field at the nucleus which usually acts in opposition to the externally applied field. This change in the effective field on the nuclear spin causes the NMR signal frequency to shift. The magnitude of the shift depends upon the type of nucleus and the details of the electron motion in the nearby atoms and molecules. It is called a "chemical shift". The precision of NMR spectroscopy allows this chemical shift to be measured, and the study of chemical shifts has produced a large store of information about the chemical bonds and the structure of molecules.

The effective magnetic field at the nucleus can be expressed in terms of the externally applied field B_0 by the expression

$$B = B_0\left(1 - \sigma\right)$$

where σ is called the shielding factor or screening factor. The factor σ is small – typically 10^{-5} for protons and $<10^{-3}$ for other nuclei (Becker).

In practice the chemical shift is usually indicated by a symbol δ which is defined in terms of a standard reference.

$$\delta = \frac{\left(v_S - v_R\right) \times 10^6}{v_R} \quad \text{quoted as ppm}$$

The signal shift is very small, parts per million, but the great precision with which frequencies can be measured permits the determination of chemical shift to three or more significant figures. The reference material is often tetramethylsilane $Si(CH_3)_4$, abbreviated TMS. Since the signal frequency is related to the shielding by

$$v = \frac{\gamma}{2\pi} B_0 (1 - \sigma) \qquad \gamma = \text{gyromagnetic ratio}$$

The chemical shift can also be expressed as

$$\delta = \frac{(\sigma_R - \sigma_S)}{1 - \sigma_R} \times 10^6 \approx (\sigma_R - \sigma_S) \times 10^6$$

π-Electron Functions

π-electrons are more polarizable than are sigma-bond electrons, as addition reactions of electrophilic reagents to alkenes testify. Therefore, we should not be surprised to find that field induced pi-electron movement produces strong secondary fields that perturb nearby nuclei. The pi-electrons associated with a benzene ring provide a striking example of this phenomenon, as shown. The electron cloud above and below the plane of the ring circulates in reaction to the external field so as to generate an opposing field at the center of the ring and a supporting field at the edge of the ring. This kind of spatial variation is called **anisotropy**, and it is common to nonspherical distributions of electrons, as are found in all the functions mentioned above. Regions in which the induced field supports or adds to the external field are said to be **deshielded**, because a slightly weaker external field will bring about resonance for nuclei in such areas. However, regions in which the induced field opposes the external field are termed **shielded** because an increase in the applied field is needed for resonance. Shielded regions are designated by a **plus sign** (+) and deshielded regions by a **negative sign(–)**.

Note that the anisotropy about the triple bond nicely accounts for the relatively high field chemical shift of ethynyl hydrogens. The shielding and deshielding regions about the carbonyl group have been described in two ways, which alternate in the display.

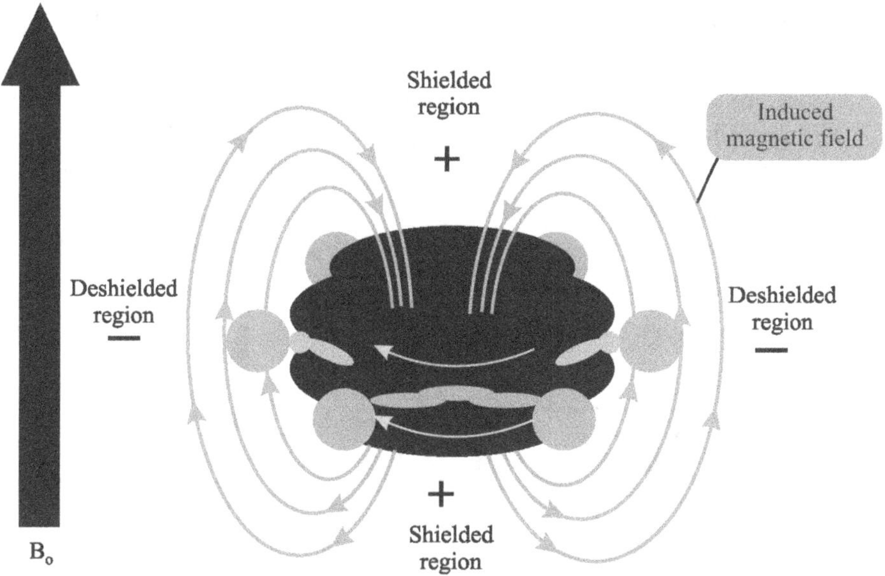

Sigma bonding electrons also have a less pronounced, but observable, anisotropic influence on nearby nuclei. This is seen in the small deshielding shift that occurs in the series CH_3–R, R–CH_2–R, R_3CH; as well as the deshielding of equatorial versus axial protons on a fixed cyclohexane ring.

Requirements and Characteristics for Spin 1/2 nuclei

- Nuclei having the same chemical shift (called **isochronous**) do not exhibit spin-splitting. They may actually be spin-coupled, but the splitting cannot be observed directly.

- Nuclei separated by three or fewer bonds (e.g., vicinal and geminal nuclei) will usually be spin-coupled and will show mutual spin-splitting of the resonance signals (same J's), provided they have different chemical shifts. Longer-range coupling may be observed in molecules having rigid configurations of atoms.

- The magnitude of the observed spin-splitting depends on many factors and is given by the coupling constant **J** (units of Hz). **J** is the same for both partners in a spin-splitting interaction and is independent of the external magnetic field strength.

- The splitting pattern of a given nucleus (or set of equivalent nuclei) can be predicted by the n+1 rule; where n is the number of neighboring spin-coupled nuclei with the same (or very similar) Js. If there are 2 neighboring, spin-coupled, nuclei, the observed signal is a triplet (2+1=3); if there are three spin-coupled neighbors, the signal is a quartet (3+1=4). In all cases the central line(s) of the splitting pattern are stronger than those on the periphery. The intensity ratio of these lines is given by the numbers in Pascal's triangle. Thus a doublet has 1:1 or equal intensities, a triplet has an intensity ratio of 1:2:1, a quartet 1:3:3:1 etc.

No coupled hydrogen:

One coupled hydrogen (1:1):

Two coupled hydrogens (1:2:1):

Three coupled hydrogens (1:3:3:1):

Chemical shift of dissolved water in deuterated solvents:

Solvent	δ
Chloroform (d$_3$)	1.5
Benzene (d$_6$)	0.4
Acetone (d$_6$)	2.75
Methylene chloride (d$_2$)	1.55
Dimethyl formamide (d$_7$)	3.0
Pyridine (d$_5$)	5.0
Toluene (d$_8$)	0.1-0.2
Methanol (d$_4$)	4.9
Acetonitrile (d$_3$)	2.1
Dimethyl sulfoxide (d$_6$)	3.35
Water (d$_2$)	4.75 (D$_2$O)

General regions of chemical shift (δ ppm):

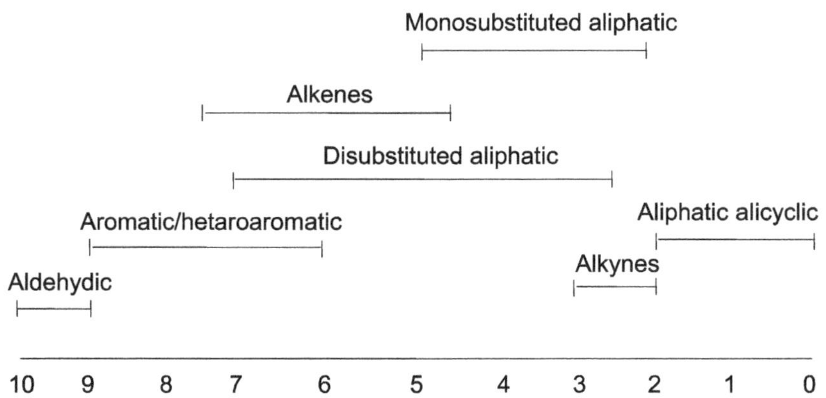

Annexure - 1

^{13}C NMR data of some natural products:

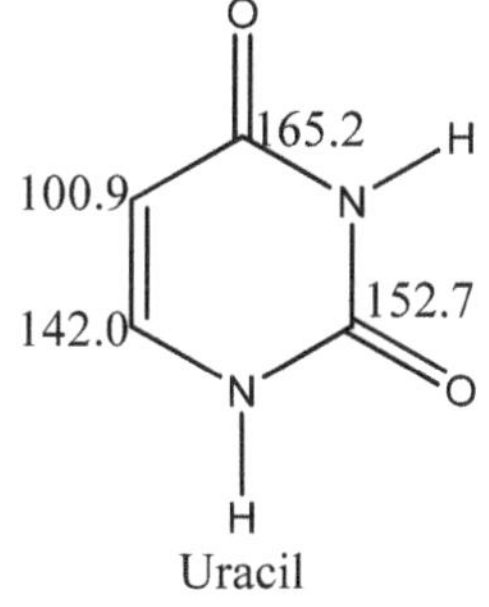

Uracil

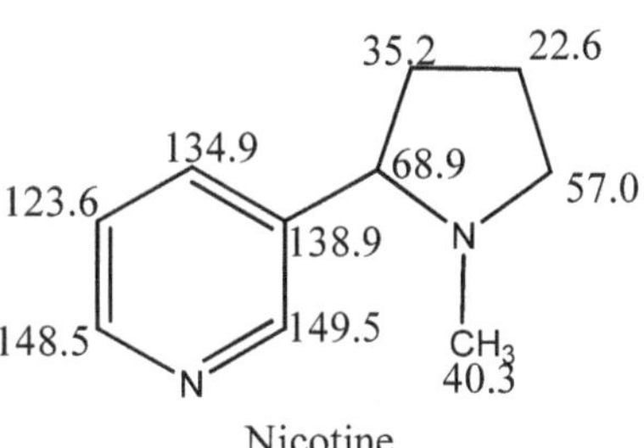

Nicotine

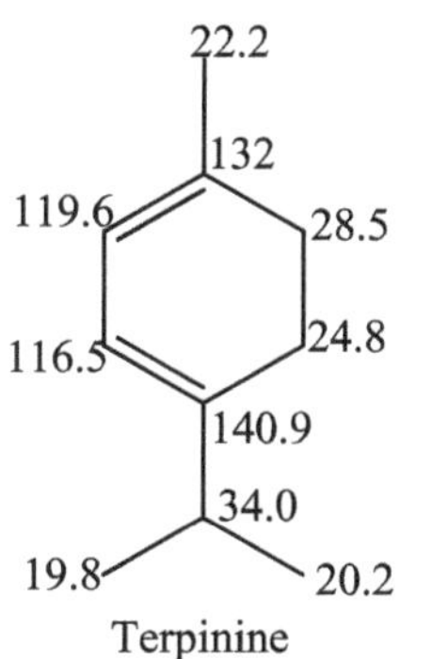

Terpinine

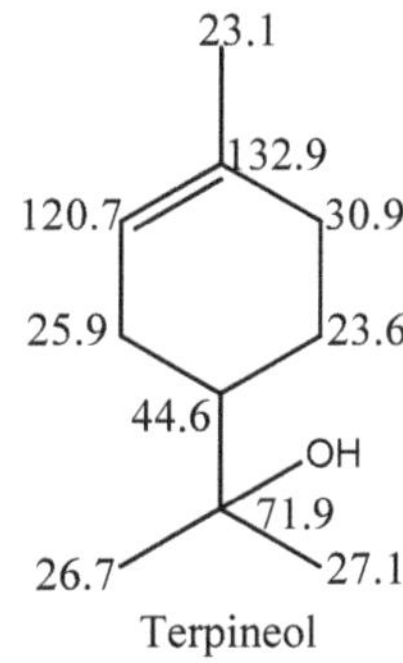

Terpineol

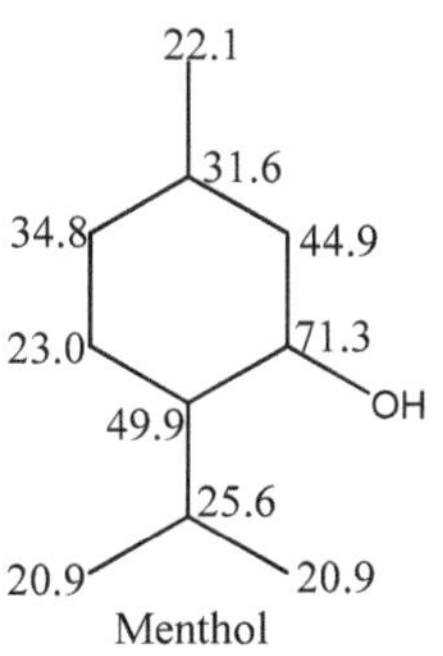

Menthol

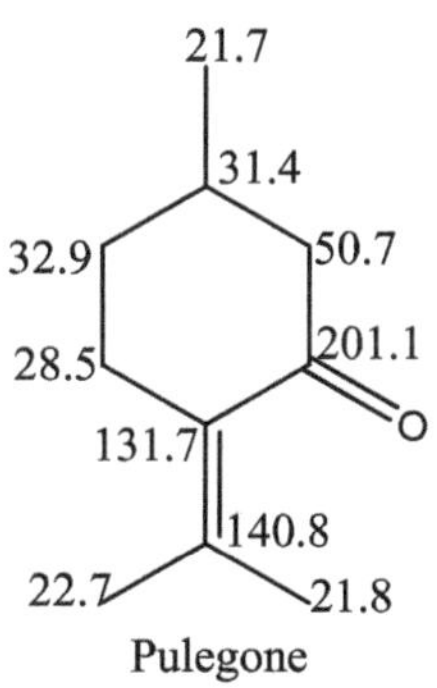

Pulegone

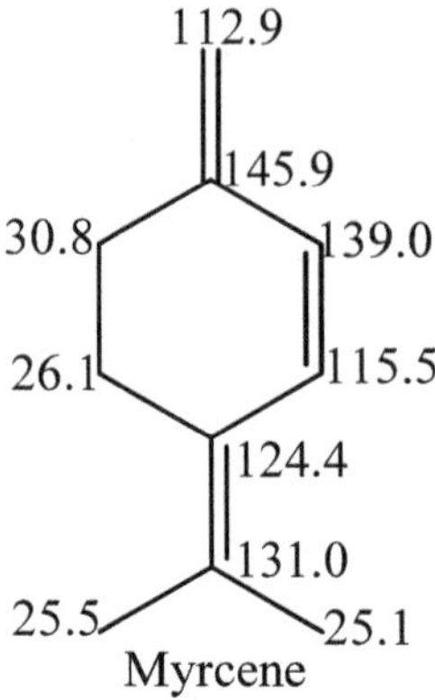

Myrcene

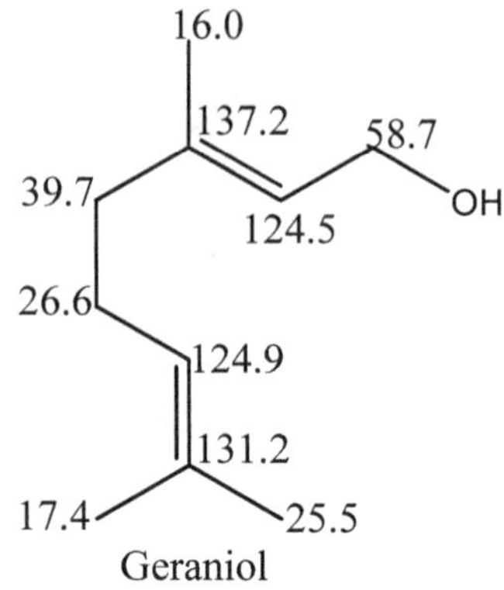

Geraniol

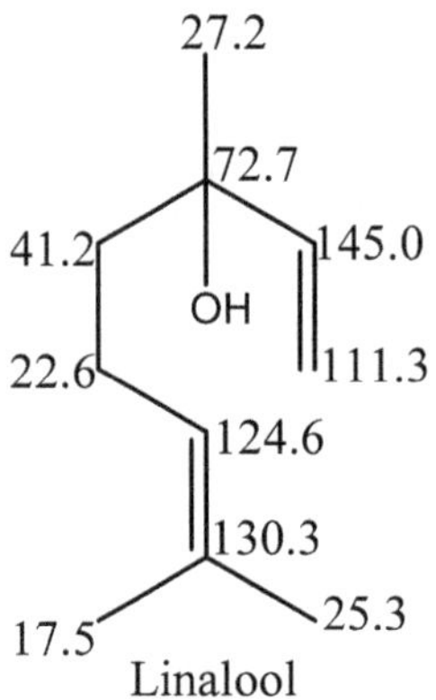

Linalool

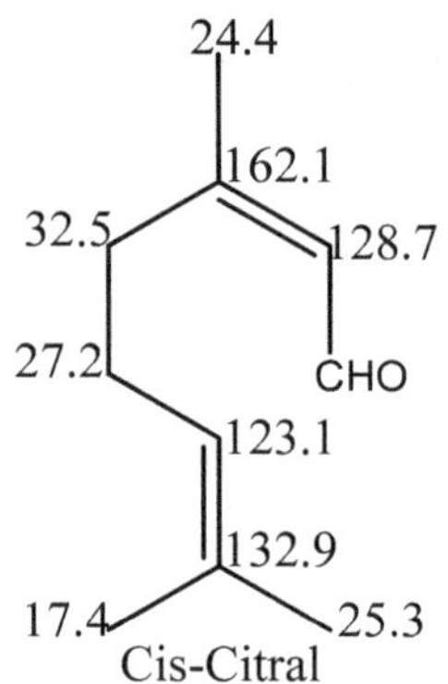

Cis-Citral

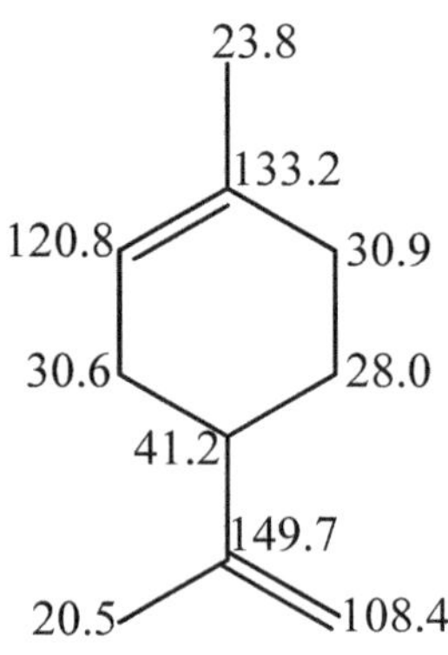

Limonene

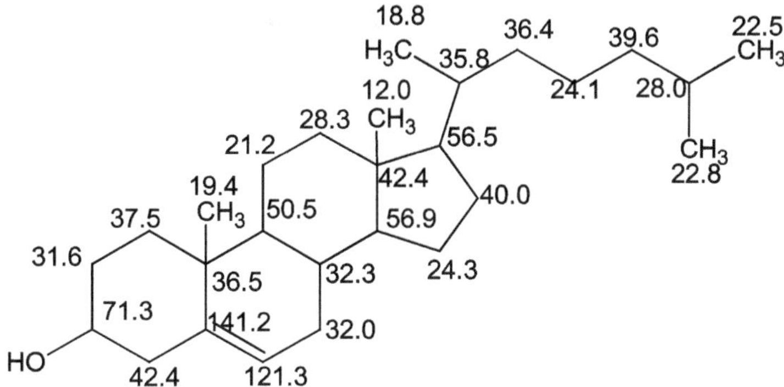

Cholesterol

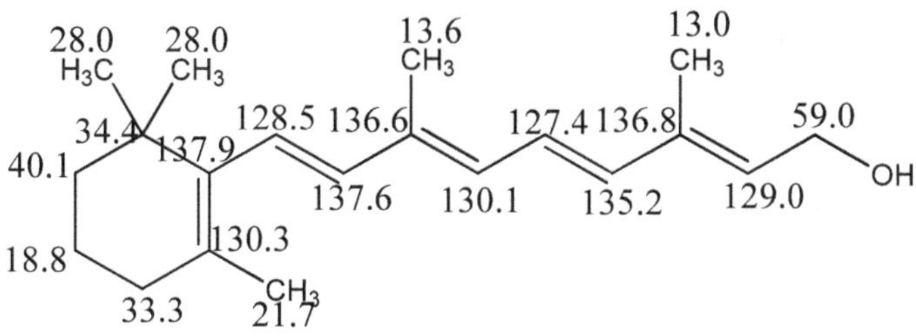

Vitamin-A

Annexure - 2

List of Common Solvents:

Solvent	B.P (^{o}C)	freezing point (^{o}C)
Diethyl ether	35	−116
Methylene dichloride	40	−97
Carbondisulfide	46	−111
Acetone	56	−95
Chloroform	61	−64
Methyl alcohol	65	−98
Tetrahydrofuran	66	--
Di isopropylether	68	−60
Carbontetra chloride	76	−23
Ethylacetate	77	−84
Ethylalcohol	78	−117
Benzene	80	5.5
Cyclohexane	81.4	6.5
Acetonitrile	82	−44
Isopropyl alcohol	82	−89
Petroleum solvents (Variable)	30 – 175	--
Water	100	0
Nitromethane	101	−29
Dioxane	102	12
Toluene	111	−95
Pyridine	115	−42
n-butyl alcohol	118	−89
acetic acid (glac)	118	17
m- xylene	139	−54

N,N-dimethylformamide	154	--
Nitrobenzene	211	5.7
Diethylene glycol	245	-10
Ethylformate	54.2	--
Methylacetate	57.1	--
Isobutylacetate	117	--
Ethylbenzoate	213	--